A Guide to the Economic Removal of Metals from Aqueous Solutions

Scrivener Publishing
3 Winter Street, Suite 3
Salem, MA 01970

Scrivener Publishing Collections Editors

James E. R. Couper	Ken Dragoon
Richard Erdlac	Rafiq Islam
Norman Lieberman	Peter Martin
W. Kent Muhlbauer	Andrew Y. C. Nee
S. A. Sherif	James G. Speight

Publishers at Scrivener
Martin Scrivener (martin@scrivenerpublishing.com)
Phillip Carmical (pcarmical@scrivenerpublishing.com)

A Guide to the Economic Removal of Metals from Aqueous Solutions

Yogesh C. Sharma

Scrivener

For general information on our other products and services or for technical support, please contact our Customer Care Department within the United States at (800) 762-2974, outside the United States at (317) 572-3993 or fax (317) 572-4002.

Wiley also publishes its books in a variety of electronic formats. Some content that appears in print may not be available in electronic formats. For more information about Wiley products, visit our web site at www.wiley.com.

For more information about Scrivener products please visit www.scrivenerpublishing.com.

Cover design by Russell Richardson

Library of Congress Cataloging-in-Publication Data:

ISBN 978-1-118-13715-4

Printed in the United States of America

10 9 8 7 6 5 4 3 2 1

Dedicated to Parents, Bebi, Kuku and Ankur!

Contents

Preface

Popular statements like 'Water is a gift of nature'; 'Water is a compound available in plenty in nature' are made every day. However, although water is plentifully available, most of it is in form of saline water contained in the oceans. Readily available freshwater is just 1% of the earth's total water mass and it is facing increasing demands and use. Industries like textile, beverages, steel manufacture, pulp and paper, and electroplating industries are among the heaviest industry users consuming enormous amounts of water. The expanding global population and the growing number of industries worldwide are escalating the demand at a rate never seen before. Because of this escalating demand of water and the fact that freshwater supply does not increase, we have to give serious thought to this problem and find solutions. One such possible solution to resolve the scarcity of water can be found in its 'treatment and reuse'.

The present book has been written the aspect of 'treatment and reuse' of polluted water in mind. Nickel and chromium are two toxic metallic species and their harmful effects are well documented. These metals have many industrial applications and their discharge of industrial effluents containing them has been identified as a main source of water pollution. There have been a number of methodologies used for treatment of metal-containing industrial effluents and waste water. Activated carbon adsorption is one of the most popular methodologies available for the purpose although it is very expensive. Silica has been a natural nontoxic substance but offers a low capacity of removal of pollutants. In the work reported in this book, modified silica has been used as an adsorbent. The process of removal has been optimized and several important studies have been reported. The problem of removal of the selected metallic species from effluents is a common problem of all nations. Developing nations like India, China, Indonesia, Sri Lanka, as well as Canada, USA and Europe will particularly benefit from reading this book. I hope this small book will serve as a guide to scientists, academics, consultants, public health workers, and environmental and water managers working in the related areas of the treatment and reuse of water.

The Author would sincerely like express his thanks to his students and others for help, advice, and friendship during the course of writing this book.

Yogesh C. Sharma

September 2011

1

Introduction

1.1 Environment

The term environment, which etymologically means surroundings, is considered a composite term for the conditions in which organisms live. Thus, it encompasses air, water, food and sunlight, which are the basic needs of all living beings and plant life necessary to carry on their life functions. Environment can also be defined as "a holistic view of the world as it functions at any time with a multitude of special elemental and socio-economic systems distinguished by quality and attributes of space and mode of behaviour of biotic and abiotic forms," or as "the sum of all social, economical, biological, physical or chemical factors which constitute the surroundings of man, who is both creator and modulator of his environment". The environment consists of the atmosphere, the hydrosphere, and the lithosphere in which the life sustaining resources of the earth are contained. The atmosphere is a mixture of gases extending outwards from the surface of earth that evolved from elements of the earth that were gasified during its

1

formation and metamorphosis. It extends to a height of about 1600 km from the earth's surface. The hydrosphere consists of the oceans, the lakes and streams, and the shallow ground water bodies. The lithosphere is the soil mantle that wraps the core of the earth. The lithosphere mainly contains three layers of crust, the mantle and outer and inner core. The biosphere, a shell that surrounds the earth, is made up of the atmosphere and lithosphere adjacent to the surface of the earth together with the hydrosphere. It is within the biosphere that the life forms of earth, including humans, live. Life sustaining materials in gaseous, liquid, and solid forms are cycled through the biosphere, providing sustenance to all living organisms.

1.2   World Water Distribution

Water is one of the most abundant compounds found in nature, covering approximately three fourths of the surface of the earth. However, despite the apparent abundance, several factors serve to limit the amount of water available for human use.

As shown in Table 1.1 [1] over 97% of the total water supply is contained in the oceans and other saline bodies of water, and is not readily available for most purposes. Out of the remaining 35%, a little over 2% is tied up in ice caps and glaciers and, along with atmospheric and soil moisture, is inaccessible. Thus, for their general livelihood and the support of their varied technical and agricultural activities, humans must depend upon the approximately 1.0% of remaining water found in fresh water lakes, rivers, and ground water supplies. Water is essential to life. Without it, the biosphere that exists on the surface of the earth would not be possible.

Water is in a constant state of motion (Figure 1). The hydrologic cycle is a conceptual model that describes the storage and movement of water between the biosphere, atmosphere, lithosphere, and hydrosphere. Water on this planet can be stored in any one of the following reservoirs: atmosphere,

Table 1.1 World water distribution.

Serial No.	Locations	Volume ($\times 10^{12}$ m³)	% of Total
(i)	Land areas		
a	Freshwater lakes	125	0.009
b	Saline lakes and inland seas	104	0.008
c	Rivers (average instantaneous volume)	1.25	0.001
d	Soil moisture	67	0.005
e	Ground water (above depth of 4,000 m)	8,350	0.610
f	Ice caps and glaciers	29,200	2.140
	Total land areas (rounded)	37,800	2.800
(II)	Atmosphere (water vapour)	13	0.001
(III)	Oceans	1,320,0000	97.300
	Total, all locations (rounded)	**1,360,0000**	**100.00**

oceans, lakes, rivers, soils, glaciers, snowfields, and groundwater. Water moves from one reservoir to another by way of processes such as evaporation, condensation, precipitation, deposition, runoff, infiltration, sublimation, transpiration, melting, and groundwater flow. The oceans supply most of the evaporated water found in the atmosphere.

Of this evaporated water, only 91% is returned to the ocean basins by way of precipitation. The remaining 9% is transported to areas over landmasses where climatological factors induce the formation of precipitation. The resulting imbalance between rates of evaporation and precipitation over land and ocean is corrected by runoff and groundwater flow to the oceans. Water as it is found in nature is almost pure in its evaporated state, contaminants are added as the liquid water

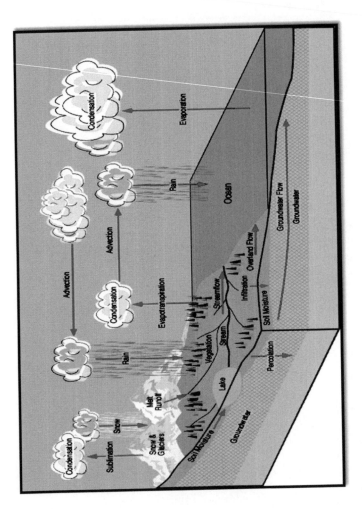

Figure 1 Hydrologic cycle.
Source: http://www.physicalgeography.net/fundamentals/5c_1.html

travels through the remainder of the hydrological cycle and comes in contact with materials in the air and on or beneath the surface of the earth. Ultimately the contaminated water will complete the hydrological cycle and will be returned back to the atmosphere again as a relatively pure water molecule. However, it is the water quality in the intermediate stage which is of greatest concern because the quality at this stage will affect human use of the water.

1.3   Environmental Pollution

From the beginning of civilization, man has always been interfering with the environment. As a consequence of the industrial revolution over the last few decades, man has had an immense impact on the environment. This impact has resulted in the deterioration of water, air, and land quality. It has also had an adverse effect on living beings due to the disturbance of the ecological balance between the living and nonliving components of earth. Due to the introduction of undesirable components into the environment, undesirable changes take place in the physical, chemical, and biological characteristics of air, water, and land, causing harmful effects on living beings.

"Environmental pollution is any discharge of material or energy into water, land, or air that causes or may cause acute or chronic detriment to the Earth's ecological balance or that lowers the quality of life." Pollutants may cause primary damage, with a direct identifiable impact on the environment, or secondary damage in the form of minor perturbations in the delicate balance of the biological food web that are detectable only over long time periods.

Environmental pollution is the build-up and concentration of toxic levels of chemicals in the air, water, and land, which reduces the ability of the affected area to support life. Pollutants may be gaseous, e.g., ozone and carbon monoxide; liquids discharged from industrial plants and sewage systems; or solid waste landfills and junkyards.

Environmental pollution can be categorized as follows:

    i. Air pollution
   ii. Soil or land pollution
  iii. Noise pollution
  iv. Water pollution

1.3.1 Air Pollution

Once substances which have been emitted exist throughout the atmosphere, over and above the physical diffusion and deposition functions, chemical elimination functions and biological purification functions, and they are dispersed with their volume exceeding natural conditions. they have a direct influence on living beings. This is called air pollution, also defined as, "the accumulation in the atmosphere of substances that, in sufficient concentrations, endanger human health or produce other measured effects on living matter and other materials".

Air pollutants can be divided in two categories, namely: primary pollutants such as carbon monoxide, hydrocarbons, nitrogen oxides, and sulfur dioxides, which are emitted directly from the sources; and secondary pollutants like photochemical oxidants, aerosols, and hydrogen chloride gas, which are created by chemical changes which occur in the atmospheric environment. Air pollution may be caused by natural sources such as volcanic eruptions, forest fires, soil debris, marsh gas, natural and inorganic decay, and cosmic dust. It may also be caused by anthropogenic sources such as deforestation, emissions from vehicles, combustion, manufacturing processes, agricultural activities, nuclear energy activities, and rapid industrialization.

1.3.2 Soil or Land Pollution

Soil/land pollution refers to the addition of solid and liquid wastes to soil, creating an imbalance in its natural composition and functions. Rapid urbanization due to an increase in population and building construction has resulted in the reduction of land. Deposition of different wastes onto land results in the deterioration of soil quality. The soil becomes polluted by hazardous chemicals resulting in microorganisms entering the food chain, air, and water to consequently be ingested by man. The main sources of this pollution are agricultural activities, industrial effluents, radioactive pollutants, solid wastes, urban activities, etc.

1.3.3   Noise Pollution

Noise pollution is a composite of sounds generated by human activities ranging from blasting stereo systems to the roar of supersonic transport jets. Although the frequency of noise may be of major importance, most noise sources are measured in terms of intensity, or strength of the sound field. The standard unit, one decibel (dB), is the amount of sound that is just audible to the average human. Noise may be generally associated with an industrial society, where heavy machinery, motor vehicles, and aircrafts have become everyday items. The most readily measurable physiological effect of noise pollution is damage to hearing, which may be temporary or permanent, and may cause the disruption of normal activities or just be a general annoyance. The effect is variable, depending upon individual susceptibility, duration of exposure, nature of noise (loudness), and time distribution of exposure such as steady or intermittent.

1.3.4   Water Pollution

The 'once-through' use of fresh water in many communities has approached or exceeded the limit of existing water supplies. With the growth of population and industries, it has become increasingly difficult to develop new sources of municipal and industrial water supplies. Water has been the most exploited natural system since the beginning of human civilization. Several industries have been set up in recent decades. This rapid technological development causes the deterioration of the water quality.

"Water pollution can be defined as any adverse change in conditions or composition of water so that it becomes less suitable or unsuitable for the purpose for which it would be suitable in its natural state".

Water is the basis of all life and is an absolute necessity for all activities, namely domestic, industrial, and agricultural. Water pollution may be caused by natural processes like decomposed vegetables, animal and weathered products, or may be caused by anthropogenic processes such as agriculture, industry, radioactivity and mining sources. Some of the major sources of water pollution are industrial effluents, agricultural wastes, atmospheric gases, radioactive materials, thermal pollutants, heavy metals, and sewage and domestic waste [2, 3]. Natural waters are contaminated with several heavy metals arising from mining wastes and industrial discharges [4].

The tremendous increase in the use of heavy metals over the past few decades has eventually resulted in an increased flux of metallic substances in the environment. Heavy metals are the metals with specific gravities of ≥ 5.0 gm cm^{-3}. Sometimes in their pure form these heavy metals are not very toxic, but when they come in contact with other components they reportedly act as very toxic pollutants. Large number of metals are commonly used in different industrial and agricultural activities. Some of them are Cr, Pb, Hg, As, Cd, Sb, Ni, Zn, Cu, and Co [5]. Chromium and nickel are widely used metals and have wide spread applications in different industries. Industrial processes that produce aqueous effluents enriched with heavy metals are shown in Table 1.2 [6, 7]. Table 1.2 also describes various metallic species and their application in industries.

It is interesting to note that most of the metals have a variety of applications. Toxic heavy metals in air, soil, and water pose a serious threat to the environment. Global discharges of some commonly used toxic metals in water, air, and soil are given in Table 1.3 [8].

It is clear from Table 1.3 that invariably the metal content is highest in soil. Like other pollutants, metallic elements are

Table 1.2 Heavy metals in some major industries.

Industry /Source	Al	Zn	As	Sn	Cd	Cr	Cu	Fe	Hg	Mn	Pb	Ni
Automobile		X		X	X	X		X			X	X
Paterolium refining		X	X			X	X	X			X	X
Pulp and paper		X				X	X		X		X	X
Textile						X						
Steel		X	X		X	X		X	X		X	X
Organic chemicals	X	X	X	X	X	X		X	X		X	
Inorganic chemicals	X	X	X		X	X		X	X		X	
Fertilizers	X	X	X		X	X	X	X	X	X	X	X
Leather tanning and finishing		X			X	X	X	X			X	X
Steel power plant		X				X			X			
Mining			X		X		X			X	X	
Acid mine drainage	X	X					X	X		X		
Metal plating		X			X	X	X					
Glass			X		X						X	
Coal and gasoline								X			X	

Table 1.3 Global discharges of trace metals (1000 metric tonnes/year).

Metals	Water	Air	Soil
Arsenic	41	19	82
Cadmium	9.4	7.4	22
Chromium	142	30	896
Copper	112	35	954
Lead	138	332	796
Mercury	4.6	3.6	8.3
Nickel	113	56	325
Selenium	41	3.8	41
Tin	ND	6.4	ND
Zinc	226	132	1372

Table 1.4 Environmental standards for metallic pollutants.

Metallic Pollutants	Threshold Limit Value (mg/L)
Cadmium	0.01
Lead	0.1
Hexavalent chromium	0.05
Nickel	0.1
Arsenic	0.05
Mercury	0.002
Selenium	0.05
Copper	0.03
Zinc	0.5
Manganese	0.05
Silver	0.05

also classified as essential and non-essential metals. Essential metals are an integral part of the diet, but non-essential metals also enter the environment and pose health problems. Due to widespread application of heavy metal in different industrial activities, higher concentrations of metallic pollutants are continuously being discharged in water bodies. The United States Environmental Protection Agency (USEPA) has decided the threshold limit value (TLV) for heavy metals in water (USEPA 1991). Environment standard for different metallic pollutants are given in Table 1.4 [9, 10].

1.4   Chromium

Chromium is a member of the transition metal series of elements in the periodic table. Its atomic number is 24 and standard atomic weight is 51.9961gmol^{-1} [11].

Chromium is an important industrial metal of interest due to its high corrosion resistance and hardness. Naturally it is distributed in rocks and soils. The common ore of chromium is chromite, $(Fe, Mg) O (Cr, Al, Fe)_2O_3$. It is the only commercially important ore of chromium [12]. About 60–70% of chromium present in the atmosphere originates from anthropogenic sources, while the remaining 30–40% is from natural sources. The natural sources of chromium are volcanic eruptions, erosion of soils and rocks, airborne sea salt particles, and smoke from forest wildfires. The anthropogenic sources contributing to the chromium increase in the atmosphere are metallurgical industries, refractory brick production, electroplating, combustion of fuels, and production of chromium chemicals, mainly chromates and dichromates, pigments and chromium trioxide. Uses of chromium and its compounds are shown in Figure 1.1.

Chromium mainly exists as metallic chromium, bivalent chromium, trivalent chromium, and hexavalent chromium. From an environmental point of view, two of its forms, namely Cr (III) and Cr (VI) are important [13]. The trivalent form is

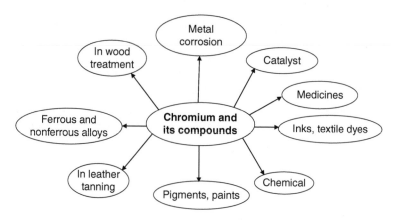

Figure 1.1 Uses of chromium and its compounds.

an essential nutrient for man and is required in amounts of 50–200 µg/day, whereas the hexavalent form is carcinogenic to man and animals. The trivalent species is more stable than the hexavalent form, and the hexavalent form is reduced to trivalent form by a variety of organic compounds. Naturally occurring chromium is almost always present in the trivalent state, whereas hexavalent chromium in the environment is derived mainly from human activities [14].

Cr (VI) compounds are highly soluble, mobile, and bioavailable as compared to the sparingly soluble trivalent species. Hexavalent chromium compounds lead to a variety of clinical problems due to their toxic nature. Inhalations of materials containing Cr (VI) are reported to cause perforation of the nasal septum, asthma, bronchitis, pneumonitis, and inflammation of the larynx, lungs, liver, and digestive system. Chromium toxicity may lead to bronchogenic carcinoma. The mechanism of the carcinogenic effect of Cr (VI) is based on the fact that Cr (VI) binds to double-stranded deoxyribonucleic acid (DNA), thus altering gene replication and duplication. Besides that, skin contact of Cr (VI) compounds can induce skin allergies, dermatitis, dermal necrosis, and dermal corrosion.

Cr (VI) compounds can also induce heritable changes in cells or organisms, and are mutagenic substances. The most important known reason for its mutagenic activity is its oxidative property. Cr (VI) ions are easily transported through the cellular membrane. Once they enter the cell, they oxidize its constituents and undergo a metabolic reduction, i.e., Cr (VI) to Cr (III). The final negative impact leads to the migration of these chromium metabolic complexes to the nucleus of the cell and their interaction with DNA [15–20].

1.4.1   Ecological Effects of Chromium

The toxicity of hexavalent chromium to aquatic species is generally low; chromium is relatively less toxic to aquatic organisms compared with metals like Cd, Cu, Pb, Hg, Ni, or Zn. It is a fact that fishes are less susceptible to the toxic effects of Cr than other organisms, especially invertebrates. The rate of uptake of Cr in fishes is high, and the Cr burden progressively declines with the age of the fishes. It is a surprising fact that Cr is not reported as a pollutant of concern in plant tissues except at the site specific discharge points! On the contrary, algae are generally susceptible to the accumulation and toxicity of chromium. Some species of algae can concentrate chromium in water up to approximately 4000 times. A concentration of 10 ppm of Cr is lethal to several species of algae. Marine plant species are reported to have more concentrations of Cr than their freshwater counterparts. This indirectly indicates an increased bioavailability of Cr to marine species.

Due to its toxic effect and carcinogenic nature in humans, the USEPA has fixed the Cr (VI) maximum contaminant level as 100 ppb for total chromium [21] and 0.05 mg/L for the domestic water supply [22, 23]. The maximum levels are 5 mg/l for trivalent chromium, and 0.05 mg/l for hexavalent chromium for wastewaters. When low levels of Cr are present in the environment, trivalent chromium apparently plays an essential role in plant and animal metabolic processes, and hexavalent Cr is directly toxic to living organisms. The chromium related industries are now facing the problem of the safest way to dispose of large quantities of chromium containing wastewater. This has become a serious issue because Cr (VI) has been classified as a Group I human carcinogen by the International Agency for Research on Cancer (IARC), and as a Group A inhalation carcinogen by the U.S. Environmental Protection Agency (EPA) [24]. The main sources of chromium pollution are leather tanning, mining, cement industries, dye industries, electroplating, steel and alloy manufacturing, photographic materials, and corrosive paints.

Chromium-laden effluents discharged from these industries need to be treated before they are released into a body of water or any adequate disposal site. There are numerous techniques for water purification and metal recovery operations from wastewaters. Many of the technologies are well established methods, while others are in early experimental stages. Practically speaking, a combination of more than one technology provides desirable results. The usual methods of metal removal from wastewaters are chemical precipitation, electrodeposition, and cementation. Other methods include activated carbon adsorption, ion exchange, and reverse osmosis, but these methods have some drawbacks. Efforts have been directed in the search for a reliable and economically suitable method for the removal of heavy metals from wastewater. The common and conventional technique frequently used by industrial treatment plants is chemical precipitation, which generally involves the reduction of Cr (VI) to Cr (III) by reductants like Fe (II), and the subsequent adjustment of the solution pH to near neutral conditions to precipitate the Cr (III) ions produced [25–27].

Depending upon the pH of the wastewater, the hexavalent chromium is primarily present in the form of chromate (CrO_4^{2-}), hydrogen chromate ($HCrO_4^-$), and dichromate ($Cr_2O_7^{2-}$) ions. At a low pH and high total chromium (VI) concentration, the dichromate ion ($Cr_2O_7^{2-}$) predominates. Dichromate only becomes important at very high Cr concentrations ($>10^{-2}$ M Cr), which might occur in cases of heavy industrial pollution. At pH > 6.5, Cr (VI) is in the form of chromate ion. The predominant Cr (VI) species that might exist in natural water are hydrochromate and chromate [12].

Apart from using the chemical precipitation technique, other techniques used for the removal of Cr (VI) from wastewater are ion exchange, solvent extraction, membrane separation, cementation, biosorption, lime coagulation, and reverse osmosis [28–30]. In the case of the chemical precipitation technique, the major disadvantage is that it requires high doses of chemicals, and during the completion of the precipitation process significant amounts of chemical sludge are produced. This chemical sludge is difficult to dewater, and further careful and frequently expensive disposal management is required. Besides, if various types of metal co-exist in the same solution or wastewater, this technique cannot be applied due to the existence of a different pH for precipitation of a given metal element [31].

Other removal techniques like ion exchange, solvent extraction, membrane separation, cementation, and reverse osmosis are costly and cannot be commercially applicable. Due to their high per capita cost, these techniques cannot be applicable in developing countries such as India, Indonesia, Sri Lanka, China, Korea, South Africa, and Brazil. Also, the problems of the removal and recovery of metals from industrial effluents in developed economies such as the United Kingdom, the United States of America, Canada, and Germany are more or less the same. Industries in these countries also tend to have low cost techniques for the removal and recovery of metals. In developing countries, pollution control measures should be based on low cost technologies. In this context, one area that is being explored is the use of naturally occurring materials that have the potential for removing pollutants by an adsorption process. The main advantage that can be gained by using adsorption is that in this technique there is no generation of undesirable chemical sludge. The adsorption technique can be applied commercially, and the cost of this technique can further be lowered by using a low cost natural adsorbent like silica sand. Many models of adsorption for cations and anions on a surface have been developed considering variations in parameters such as pH, adsorbent dose, adsorbate concentration, time, and even ionic strength. The main applications of these models are to predict the optimum adsorption conditions for a given adsorbent with adsorbate [32].

Adsorption when combined with the appropriate step of desorbing the Cr (VI) from the adsorbent, and which avoids the problem of disposal, is a cost effective and versatile method for the removal of Cr (VI) [33]. Other advantages of the adsorption process are that it can be exercised with other materials having the same structural, compositional, or chemical characteristics suitable to endow this technique with high Cr (VI) retention values. Thus, it has high potential for the removal of Cr (VI) from wastewater streams [34].

The selection of an adsorbent is a key factor for the use of adsorption in the treatment technique for Cr (VI) removal. The cost associated with commercial adsorbents make the adsorption process very expensive. This has led to the search for new strategies for developing low-cost adsorbent materials with a good capacity for Cr (VI) removal.

Nowadays, the adsorption process for chromium removal is being carried out by using naturally available biomaterials such as seaweed, fungal biomass, green algae, maple sawdust, soya cake, red mud, Turkish brown coal, moss peat, hazelnut shells, coconut trees, lignocelluloses residues, rice bran, activated neem leaves, activated tamarind seeds, Bengal gram husk, eucalyptus bark, sawdust, sugarcane bagasse, sugar beet pulp, coconut husk fibers, palm pressed fibers, waste tea, Ocimum basilicum seeds, cocoa shell, biogas residual slurry, and zeolite [35, 36]. Several reports reflect that sand is an inexpensive and efficient adsorbent for the removal of heavy metal from industrial effluents [37].

1.5   Nickel

Nickel is the 24th element in order of natural abundance in the earth's crust. Nickel occurs naturally in the form of ores such as oxides, silicates, and sulphides, and is usually associated with other sulphide, silicate, or arsenide minerals [38]. Forty-seven percent of nickel is used in steel production (steels with increased strength and resistance to corrosion and temperature), 21% in the production of other alloys, and 12% as nickel (II) sulphate and nickel (II) chloride in electroplating. Nickel is also used in coinage, as a catalyst (as nickel oxide), in ceramics, in storage batteries (nickel-cadmium batteries) as nickel hydroxide, in a dying process for polypropylene, for colouring glass, in electronic components (as nickel carbonate), in food processing equipment, and in certain fungicides because fungi are sensitive to nickel. It is also used in the production of stainless steel, and common brands of stainless steel contain 8% Ni and 18% Cr. Figure 1.2 depicts the various applications of nickel.

Natural sources of nickel include continental windblown dust, volcanic dust and gases. The burning of residual and fuel oils, the mining and refining of nickel, and municipal waste incineration are the major anthropogenic sources of nickel emissions to the atmosphere. Total annual nickel emissions from anthropogenic sources have been calculated to be 9.8×10^{10}g, whereas emissions from natural sources contribute 3×10^{10}g per year to global atmospheric emissions. Nickel exists in many valence states such as metallic nickel (Ni^0), monovalent nickel (Ni^{+1}), and trivalent nickel (Ni^{+3}).

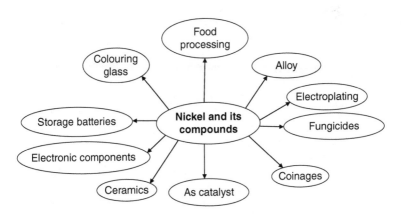

Figure 1.2 Uses of nickel and its compounds.

From an environmental point of view only one form of nickel (i.e., Ni^{+2}) is important [39]. Emission of Ni from fossil fuels is 70,000 tonnes/year. It should be noted that city air is contaminated with 0.03–0.12 g/m^3 of Ni, and this amounts to an approximate daily intake of 0.30–1.2 g for humans.

Though nickel is insoluble in water, many nickel salts are readily soluble in water. Hence, water contamination by nickel is common. Nickel is known to cross the animal and human placental barrier. Due to its deep reach inside the human body, it can induce embryotoxic and nephrotoxic effects, allergic reactions, and contact dermatitis. Nickel alloys and nickel compounds are among the most common causes of allergic contact dermatitis. Nickel sensitization occurs in the general population from exposure to nickel-containing coins, jewelry, watch cases, clothing fasteners, and other objects, and results in conjunctivitis, eosinophilic pneumonitis, and asthma. Nickel sensitization also occurs due to local or systemic reactions to nickel-containing prostheses such as joint replacements, intraosseous pins, cardiac valve replacements, cardiac pacemaker wires, and dental inlays. Nickel in the ionic form can cross the cell membrane and get deposited in the nucleus and the nucleolus. Thus, nickel might be both an initiator and promoter of cancer. The erythropoietic stimulation exerted by nickel is correlated with its carcinogenicity [39, 40]. Nickel is an essential micronutrient for microorganisms and animals, but it is nonessential for plants. It is reported to be associated with the synthesis of vitamin B_{12}, but is toxic at higher concentrations. In animals, the toxic effects include many skin diseases and respiratory disorders. The dermatitis from exposure to nickel is widespread among the general population, and most of the susceptible people are women using nickel containing jewelry. Nickel is reported to inhibit enzymes such as maleic dehydrogenase, cytochrome oxidase, and isocitrate dehydrogenase. Nickel carbonyl, a volatile compound formed by the combination of Ni and CO is reported to be the most toxic of all nickel compounds. It causes cancer, and a 30 minute exposure to a 30 ppm concentration is lethal to humans.

Nickel is found in various foodstuffs, but it is mainly found in tea, cocoa, and peanuts. Dietary intake of this metal is a result of the use of cooking utensils made of stainless steel. The extent of the dietary intake averages 300–600 μg/day.

Due to the toxic and carcinogenic nature of Ni (II), the World Health Organization (WHO) has set a permissible limit of Ni (II) for drinking water to be at 0.02 mg/L [41]. During the last 50 years, ample amounts of nickel have been liberated to the soil environment, especially by the deposition of ash residues from coal combustion and the disposal of municipal sewage sludge. Therefore, there is a tremendous need to treat these wastes using suitable techniques before disposal. The conventional methods for heavy metal removal from water and wastewater include oxidation, reduction, precipitation, membrane filtration, ion exchange, and adsorption. Among all of these, adsorption is highly effective and economical. Specific sorption of Ni^{2+} to mineral surfaces is strongly pH-dependent; with increasing sorption as deprotonisation of reactive sites takes place. The pH of the industrial effluent determines the type of Ni species found in the wastewater and thus the adsorbent required for removal of that Ni species. Over most of the pH range, Ni^{2+} and $NiSO_4^0$ are the predominant species. $Ni(OH)_2^0$ and $Ni(OH)_3^-$ appear only when the pH is highly basic. In the pH range of 8–10, $NiOH^+$ and $NiHCO_3^+$ may also be present [41, 42].

The adsorption technique has many advantages over other removal techniques. It is a low cost removal technique, and the cost can be lowered further by using naturally available adsorbent. In the adsorption technique, the adsorbent can further be regenerated by a suitable desorption process, and the used adsorbent can be recycled further resulting in the cost of fresh adsorbent being saved. The adsorption process does not create undesirable sludge, which occurs in the chemical precipitation process.

In my project work silica sand is the adsorbent used for the removal of Cr (VI) and Ni (II) from aqueous solutions by the batch mode of the adsorption process. Silica sand is a natural material found in riverbeds in abundance. The objective of my project work is to explore the possibility of using silica sand as an adsorbent to purify the real industrial wastewaters ladened with heavy metal contamination. The most important part of this study is to investigate the removal of Cr (VI) and Ni (II) by modified silica sand (with 40% H_2SO_4) as an alternative media, and to determine the factors affecting the adsorption of these heavy metals on silica sand. In addition, the sorption capacity, kinetics, and thermodynamics will also be determined.

1.6   Objectives

The main objective of the present study is to establish an economically viable technology for the removal of Cr and Ni from aqueous solutions. The pointwise objectives, however, are as follows:

    i. Characterization of silica sand.
   ii. Processing/Pretreatment of the adsorbent.
  iii. Use of adsorbent for the removal of Cr and Ni from the aqueous solution.
  iv. Optimization of process parameters for the removal process.

1.7   Literature Review

1.7.1   Water Pollution by Heavy Metals and the Removal of Nickel and Chromium

Heavy metals, though essential for industrial development, have also been recognized as major pollutants for fauna, flora, and humans. The presence of heavy metal contaminants in aqueous solutions, arising from the discharge of untreated metal containing effluents into water bodies, is one of the most important environmental issues. The major source of metallic pollutants in aquatic systems is the discharge of untreated industrial effluents from different industries such as electroplating, dyeing, battery manufacturing, mining operations, chemical manufacturing, tanneries, glass manufacturing and pharmaceutical manufacturing [43–46].

The metals, namely chromium, cadmium, arsenic, nickel, zinc, copper, lead, and mercury, are regularly used in various industries. Their toxicity is well documented [44]. The heavy metals are of special concern because they are nondegradable and therefore persistent in the environment [47, 48]. They have a tendency to enter into living tissues. These metals are reported to be harmful even at low concentrations. Heavy metals have a great affinity to attack sulphur bonds, protein, carboxylic acid, and the amino group, thereby disrupting the cell's metabolism. An excessive presence of these metals in living species results in carcinogenic, mutagenic, teratogenic, and other toxic effects on them. Once they are accumulated in living tissues, they disturb microbial processes and have been reported to be fatal [49, 50]. A slight excess of metals in water is extremely toxic to fish and can also damage algal growth. Chromium and nickel are the most commonly used metals in different industrial activities. Chromium, nickel, and their compounds are widely used in electroplating, dye industries, leather tanning, cement industries, and photographic industries [51, 52].

Copper, cadmium, and lead are some other important metals which are frequently used in electroplating, metal finishing, electrical industries, fertilizers, and wood manufacturing. They are also used in pigment industries as alloys with iron and other metals, and in the petrochemical and paint industries. These metals have a toxic effect at their higher concentration. Metals are considered to be indestructible poisons [53]. Their damage done to seas, rivers, and streams over a long period of time may be highly dangerous because they may affect the production of atmospheric oxygen, contaminate the water, and affect aquatic life. Thus, industrial effluents must be treated prior to their discharge into water resources.

Removal of Cr (VI) from industrial effluents by adsorption on an indigenous low-cost clay mineral called wollastonite has been reported by Sharma *et al.* [54, 55]. He concluded that the removal is dependent on the concentration of Cr (VI), and low concentrations favor the uptake. The uptake increased from 41.7–69.5% by decreasing the concentration of Cr (VI) in solution from 2.0×10^{-4} to 0.5×10^{-4}M at 0.01M $NaClO_4$ ionic strength, 2.5pH, and at 30°C. The higher temperature favors the uptake, i.e., endothermic nature of the process, and removal is higher in the lower pH range, maximum (69.5%) at 2.5pH and minimum (12.3%) at pH 8.0.

Removal of Cr (VI) from contaminated water by using chitosan and waste slurry from industry has been reported by Babel *et al.* [56]. Their findings reveal that adsorption capacity for removal of Cr (VI) are 273 mg/g at pH 4.0 and 640 mg/g for chitosan and waste slurry respectively.

Removal of Cr (VI) from aqueous solutions by different types of sand (white, yellow, and red UAE sand) was also a low cost abundant adsorbent [57]. White, yellow, and red sand from the United Arab Emirates (UAE) were employed at various adsorbent/metal ion ratios. The effect of contact time, pH, temperature, metal concentration, and sand dosage was studied. The optimum pH for adsorption was calculated as 2.0 for Cr (VI). The optimal adsorption time for Cr (III) ions was recorded as 3 hr. Even at the optimal pH, adsorption of Cr (VI) on all sand forms was very low (removal <10%) and could not be fitted to any of the common isotherms. While at pH 5.0 Cr (VI) was not adsorbed at all.

Removal of Cr (VI) ions from aqueous solution under different conditions was investigated using activated alumina (AA) and activated charcoal (AC) as adsorbents by Mor *et al.* [58]. Batch mode experiments were conducted to study the effects of adsorbent dose, contact time, pH, temperature, and initial concentration of Cr (VI). Results showed that the adsorption of Cr (VI) depended significantly on the pH and temperature. Equilibrium studies showed that Cr (VI) had a high affinity for AA at pH 4 and AC at pH 2. For AA, maximum adsorption was found at 25°C, indicating exothermic adsorption, while for AC, maximum adsorption was at 40°C.

Application of riverbed sand, a non-toxic substance for the removal of Cr (VI) from aqueous solutions has been investigated by Y.C. Sharma *et al.* [59]. Removal of Cr (VI) was dependent on the initial concentration, and removal increased from 43.2–74.3% by decreasing the initial concentration from 7.5×10^{-5} M to 1.0×10^{-5} M at 25°C, 1.0×10^{-2} M NaClO$_4$ ionic strength, and 100 rpm. Higher removal was obtained with particles of smaller sizes of the adsorbent. Removal decreased from 74.3–40.7% by increasing the temperature from 25°–35°C exhibiting exothermic nature of the process of removal. Thermodynamic parameters, namely those that change in free energy (ΔG^0), enthalpy (ΔH^0) and entropy (ΔS^0) were calculated and found to be –0.81 Kcal mol^{-1}, –17.21 Kcal mol^{-1}, and 56.94 Cal mol^{-1}, respectively; at 25°C. pH of the solution they have a pronounced effect on the removal, and a higher removal was obtained in acidic pH range, maximum (74.3%) being at 2.5 pH.

Yeast biomass for biosorbing chromium metal ion in aqueous solutions was reported by Lokeshwari *et al.* [60]. Experiments were carried out as a function of pH, temperature, biosorbent concentration, chromium concentration, contract time, agitation speed, interference, and reusability. A removal of 95% was achieved under optimized conditions. The mechanism of metal sorption by yeast cells gave good fits for Freundlich and Langmuir models with max. q value of 86.95 mg/g. The change in entropy (ΔS^0) and the heat of biosorption (ΔH^0) of biomass were estimated as 105.1 Jmol^{-1}K–1, and 48865 J mol^{-1} respectively. The free energy change (ΔG^0) obtained for the biosorption process at 31°C, initial chromium concern of 120 mg/L, and pH at 7 is –16.31 KJmol^{-1}. The high negative value of change in Gibbs free energy indicates that the spontaneity and feasibility of yeast biomass has potential for the uptake of chromium from industrial effluents.

Investigation on hexavalent chromium removal from aqueous solution by adsorption on treated S. Robusta (sal) sawdust was reported on by Baral *et al.* [61]. The studies were conducted by varying various parameters such as contact time, pH amount of adsorbent, concentration of adsorbate, and temperature. The kinetics of adsorption of Cr (VI) ion followed pseudo second order. The adsorption data fitted well to the Langmuir adsorption isotherm. The equilibrium time is 3 hr for the adsorbent having a concentration of 30 mgL^{-1}. The percentage of adsorption increased with decrease in pH, and showed maximum removal of Cr (VI) in the pH range of 4.5–6.5 for initial concentration of 5mgL^{-1}.

Removal of hexavalent chromium from aqueous solutions by Ocimum americanum L. seed pods was investigated by Levankumar *et al.* [62]. The optimum pH and agitation speed were found to be 1.5 and 121 rpm. The equilibrium adsorption data fit well with the Langmuir isotherm. The maximum chromium adsorption capacity determined from the Langmuir isotherm was 83.33 mg/g dry weight of seed pods at pH 1.5 and shaker speed 121 rpm. The batch experiments were conducted to study the adsorption kinetics of chromium removal for the concentrations of 100 mg/L, 150 mg/L, and 200 mg/L. chromium solutions. The adsorbent dosage was 8 g dry seed pods/L. The removal efficiency observed for all the three chromium concentrations was 100%. The equilibrium was achieved in less than 120 min for all three concentrations.

Utilization of waste material such as sawdust for the removal of Cr (VI) from the industrial wastewater streams was investigated [63, 67]. The maximum percentage removal of Cr (VI) is obtained at pH 1. The dominant form of Cr (VI) is HCrO$_4^-$ and the surface of adsorbent is charged positively at lower pH values. It is found from the kinetic studies that the adsorption rate of Cr (VI) is faster for the initial 250 min and then it decreases in the later part of adsorption. The equilibrium time for Cr (VI) adsorption on sawdust is obtained as 1050 min.

The equilibrium data are best fitted with the Langmuir isotherm model which confirms the monolayer adsorption of Cr (VI) onto the sawdust. The maximum adsorption capacity is obtained with the application of the Langmuir isotherm model as 41.52 mg/g, which is a comparatively good adsorption capacity. The saturated adsorption is regenerated by an acid and base treatment. The adsorption efficiency of the regenerated sawdust is found to be more than 95% of fresh sawdust for the removal of Cr (VI). The possible solution for the major problem of the desorption process [i.e., the disposal of the acid and base solution obtained that contains a high concentration of Cr (VI)] is the precipitation of Cr (VI) as Barium Chromate.

Table 1.5 shows various low-cost commercial adsorbents and their adsorption capacity, which are used for removal of Cr (VI) from aqueous solutions.

Table 1.5 Maximum adsorption capacity of various commercial and low-cost adsorbents for removal of Cr (VI) from aqueous solution [28].

S.No.	Adsorbent	Maximum Adsorbent Capacity, qm (mg/g)	Optimum pH
1	Activated neem leaves	62.97	2
2	Activated carbon (Filtrasorb – 400)	57.70	–
3	Bentonite clay	49.75	3.0
4	Sawdust	41.52	1.0
5	Activated tamarind seeds	29.70	2.0
6	Coconut husk fiber	29	2.05
7	Palm pressed-fibers	15.0	2.0
8	Tea factory waste	27.24	2.0

(Continued)

Table 1.5 (cont.) Maximum adsorption capacity of various commercial and low-cost adsorbents for removal of Cr (VI) from aqueous solution [28].

S.No.	Adsorbent	Maximum Adsorbent Capacity, qm (mg/g)	Optimum pH
9	Leaf mould	25.9	2.5
10	Pine needles	21.50	2.0
11	Almond	10	2.0
12	Cactus	7.08	2.0
13	Coal	6.78	2.0
14	Coconut shell based activated carbon	20	2.5
15	Sugar beet pulp	17.2	2.0
16	Maize cob	13.8	1.5
17	Sugar cane bagasse	13.4	2.0
18	Activated charcoal	12.87	2.0
19	Activated alumina	7.44	2.0
20	Polymer grafted sawdust	9.4	3.0
21	Maple sawdust	8.2	4.0
22	Biomass residual slurry	5.87	2.0
23	Fe(III)/Cr(III) hydroxide	1.43	5.7
24	Distillery sludge	5.7	2.5
25	Calcined bauxite	2.02	3.8
26	Fly ash impregnated with aluminium	1.8	2.0

(Continued)

Table 1.5 (cont.) Maximum adsorption capacity of various commercial and low-cost adsorbents for removal of Cr (VI) from aqueous solution [28].

S.No.	Adsorbent	Maximum Adsorbent Capacity, qm (mg/g)	Optimum pH
27	Waste tea	1.55	–
28	Walnut shell	1.33	–
29	Agricultural waste biomass	0.82	2.0
30	Rice husks	0.6	–
31	Soya cake	0.28	1.0
32	Riverbed sand	0.15	2.5

The sorbent behaviour of (natural and exfoliated) vermiculite minerals was studied with respect to metal cation Ni (II) by by Ayuso and Sanchez [64]. The Langmuir model was found to describe the sorption process well, showing the maximum sorption capacities of natural vermiculite to be high enough (b_{Ni} = 19.3 mg g^{-1}) for its use in metal waste purification systems to be feasible. These positive results are ratified when real wastewaters are used at lab-scale as well as semi-industrial scale, validating the use of this mineral as a cost-effective treatment to purify such wastewaters.

Sorption behaviour of Ni(II) was investigated by Petruzzelli *et al.* [65] on a sandy loam soil with and without a sewage sludge extract (CO_2-H_2O solution at pH 3.8), to evaluate the effects of ligand species present in the sludge in the process of retaining heavy metals. The addition of sewage sludge extract did not reduce the sorption, but rather increased the amount of heavy metals retained by soil. With untreated soil 47–66% of the added metal was sorbed, while in the presence of sludge extract an increase in sorption to a range between 81–70% was found.

The influence from 0.5 and 5 mmol L^{-1} citrate and arginine on nickel sorption in the sandy loamy topsoil of a Mollic Hapludalf has been investigated by Poulsen et al. [66] in order to evaluate the effect of organic complexing agents on nickel retention in soils. In the pH range 4.5–8, sorption edges showed a dramatic decrease in nickel sorption in the presence of citrate, but not in the presence of arginine. These differences were further quantified at pH 6.0 through sorption isotherms covering initial nickel concentration in the range of 3.4–42.6µmol L^{-1}. Again, the amounts of nickel sorbed without and in the presence of 0.5 and 5mmol L^{-1} arginine were not significantly different, whereas 0.5 and 5mmol L^{-1} citrate reduced sorption by 50–90%. The results indicate that trivalent positively charged nickel-arginine complexes ($NiH_3arg_2^{3+}$) are sorbed to cation exchange sites, while in the presence of citrate sites for the bonding of monovalent negatively charged nickel-citrate complexes, ($Nicit^-$) are sparse.

A study was conducted by Yabe et al. [67] to evaluate the removal of heavy metals in an aquatic system without prior treatment using solid adsorbents such as sand, silica, coal, and alumina. An important aspect of the proposed method was that the removal was performed at a pH range in which a given metal undergoes an adsorption process, making the method useful for wastewater treatment. After the adsorption process, removal was observed to be approximately above 80% for Ni, using alumina as an adsorbent.

Natural iron oxide-coated sand (NCS), extracted from the iron ore located northwest of Tunisia, was employed to investigate its capacity to remove nickel from aqueous solutions, and satisfactory results were reported [68]. The effects of agitation time, pH, initial metal ion concentration, and temperature on the removal of these metals were studied. In order to study the sorption isotherm, two equilibrium models, the Freundlich and Langmuir isotherms, were analyzed. The effect of a pH solution on the adsorption onto NCS was studied in the pH range from 2–9. The adsorption was endothermic, and the computation of the parameters, ΔH, ΔS, and

ΔG, indicated that the interactions were thermodynamically favourable.

The batch removal of divalent nickel (Ni^{+2}) from wastewater under different experimental conditions using economic adsorbents was investigated [69]. These natural adsorbents were from different sources, i.e., starch, activated charcoal, wood charcoal, and clay (type bleaching earth). The present work contains the effect of pH and concentration of biomass on the adsorption kinetics. The batch sorption kinetics was studied for first order reversible reaction, pseudo first, and second order reaction. The result indicates that the optimum pH for the removal of nickel (II) ions by starch, wood charcoal, activated charcoal, and clay is 7.5. The clay shows good results for removing nickel ions. The sorption of this metal over sorbent is endothermic. The pseudo-first-order chemical reaction model provides the best correlation of the data.

The main parameters influencing Nickel (II) metal sorption on maple sawdust was studied by Shukla *et al.* [70]. The parameters studied were: initial concentration, amount of adsorbent, and pH value of solution. The maximum percent metal removal was attained after about 1 h. The greatest increase in the rate of adsorption of metal ions on sawdust was observed for pH changes from 2–5. High removal has been recorded at low surface loading. Maximum percent removal of nickel (II) was found to be at pH 9.0, while the minimum is at pH 2.0.

A study was conducted by Li *et al.* [71] to examine the effectiveness of 4.0–4.75 mm crushed shells and *Sphagnum* peat moss as low-cost natural adsorbent filter materials for the removal of nickel ions from binary aqueous solutions. Metal removal efficiencies and adsorption capacities for each of the columns were estimated to identify the better filter material and operational conditions for the treatment of nickel. During the column testing, a flow rate of 1.5 mL/min (surface loading of 27.5 cm^3/cm^2 day), and bed depth of 15cm were found to represent the better operational conditions, where 42.7% nickel removal was obtained under these operational conditions.

Removal of Ni^{+2} by adsorption in a fixed bed of wheat straw was investigated by Doan *et al.* [72]. Ni^{+2} in a solution were removed by biosorption in a fixed bed of wheat straw (*Triticum aestivum*). The effect of the solution pH, temperature, and particle size (0.5, 1.0, 1.5, and 2.0 in) on biosorption of Ni^{+2} was also investigated. Biosorption of Ni^{+2} increased significantly with the solution pH from 4.0–7.0. On the other hand, removal appeared to be insensitive to liquid temperature from 25°–30°C. Nevertheless, a 25% increase in the percentage removal of metal ions was observed with further increase of liquid temperature from 30° – 35°C. However, the biosorbent particle size did not seem to have a systematic effect on the biosorption of Ni^{+2}.

Table 1.6 shows the natural materials which are used as an adsorbent for the removal of Ni (II) from aqueous solutions.

Table 1.6 Natural materials used as adsorbents for removal of Ni (II).

S.No.	Adsorbent Used	References
1.	Vermiculite minerals (natural and exfoliated)	[46]
2.	Sandy loam soil	[47]
3.	Sandy loam topsoil	[48]
4.	Sand, silica, coal, and alumina	[49]
5.	Natural iron oxide-coated sand	[50]
6.	Starch, activated charcoal, wood charcoal, and clay	[51]
7.	Maple sawdust	[52]
8.	Cushed shells and *Sphagnum* peat moss	[53]
9.	Wheat straw	[54]

1.8   Adsorption

Adsorption was first observed by C. W. Scheele in 1773 for gases, followed by Lowitz's experiments in 1785 for solutions [73]. Adsorption is the accumulation of a substance at a surface or the interface between two phases. Adsorption is a surface phenomenon. The adsorption process can occur at any interface such as liquid-liquid, gas-solid, gas-liquid, or liquid-solid, but liquid-solid interface is of special interest from the water and wastewater treatment point of view.

Adsorption can be differentiated on the basis of the forces which are operative during the process.

1.8.1   Physical Adsorption

Physical adsorption is reversible and rapid. Molecules adsorbed by physical adsorption are held to the surface by weak van der Waals forces of attraction. Adsorbed molecules are not fixed to a site at the surface, but rather are free to undergo translational movement within the interface. Physical adsorption is usually predominant at low temperature and is characterized by a relatively low energy of adsorption. Activation energy involved in cases of physical adsorption is 5–10 Kcal/mol.

1.8.2   Chemical Adsorption

Chemical adsorption occurs as a result of chemical reaction between the adsorbate molecules and the adsorbent. Various types of functional groups are present at these sites which participate in adsorption through electron sharing reactions. Chemically adsorbed molecules are considered not free to move on the surface or within the interface. Chemical adsorption is favoured by a higher temperature because chemical reaction proceeds more rapidly at elevated temperatures. In chemical adsorption, activation energy is in the range of 15–30 Kcal/mol.

1.8.3   Adsorption from Solutions

A solid surface in contact with a solution has the tendency to accumulate a surface layer of solute molecules due to an imbalance of surface forces. A majority of the molecules accumulated at the interface are adsorbed onto the large surface area within the pores of adsorbents, and relatively few are adsorbed on the outside surface of the particle. The affinity of adsorbed molecules for adsorbent varies with force field at liquid-solid interface. Adsorption of solutes from a solution is influenced largely by the competition between the solute and solvent molecules for adsorption sites.

The tendency of a particular solute to get adsorbed is determined by the difference in the adsorption potential between the solute and the solvent. When the solute-solvent affinity is large, the ability of the adsorbent to adsorb solute is usually limited. In general, the lower the affinity of the adsorbent for the solute, the higher will be the adsorption capacity. Adsorbents such as activated carbons and polymeric adsorbents, have high adsorption capacities in water because a low adsorption potential energy is needed on the part of an adsorbate to displace water from the surface of these adsorbents.

The affinity of the solute for the surface must also be larger than the affinity of the solute for the solvent in order for adsorption to be highly favoured. The solubility is a direct measure of the affinity of the solute for the solvent, and an inverse relationship between the adsorbate solubility and adsorption capacity has been observed. Less soluble solutes tend to adsorb more readily, and more soluble ones tend to be adsorbed less readily.

Adsorption from solutions is generally limited to monolayer coverage on the adsorbent surface. The adsorptive forces are weak beyond the first monolayer. The mutual attraction of solutes in the first monolayer for unadsorbed solute molecules can be assumed to be equal to the attraction of a surface of pure liquid solute for dissolved solute molecules. However, the pure liquid solute will be spontaneously dissolved at any concentration. Therefore, adsorption from solution beyond the first monolayer rarely occurs. During the transfer of adsorbate from solution to the adsorbent's sites, the high energy sites are occupied instantaneously and subsequent adsorption process, as the surface coverage increases, takes place at low energy level sites, i.e., at less favourable sites. In this situation, the heat of adsorption and effective diffusion rate decline rapidly with an increasing surface coverage. This preferential trend continues until the solute phase concentration remaining in the solution is in equilibrium with the adsorbed phase concentration.

When dynamic equilibrium is established, the transfer of molecular species no longer exists, and thus molecular distribution between the solution and the solid phase is measurable and well defined. The equilibrium data thus generated can be used to test the isotherm. The rate data is usually analysed using diffusion models to justify the advanced steps of intraparticle mass transport which determines the rate of reaction, size, and structure of individual ions or solute molecules. Thus, equilibrium and distribution of adsorbate molecules between two phases are important properties of adsorption which help in defining the capacity of a particular system. The approach to equilibrium is very rapid for a surface to which adsorbate molecules have easy access. The approach to equilibrium is caused by the slow transfer of the adsorbate molecules from the bulk solution surrounding the particles to the active adsorption sites inside it. Nearly all the adsorption sites exist in the interior of the adsorbent.

Before adsorption starts, the adsorbate molecules are dispersed in the bulk liquid. The rate of adsorption is determined by the rate of transfer of the adsorbate from bulk solution to the adsorption sites within the particles. In order to get adsorbed at a particular site, the adsorbate must diffuse across a stationary solvent film surrounding each adsorbent particle through the macro pores, and then through the micro pores. The time for adsorption to take place is the sum of the time taken for each step. The slowest step will determine the rate for the whole process, and is called the rate determining step.

1.8.4 Film Diffusion

The random movement of adsorbate molecules resulting from the collisions with the solvent molecules lead to their diffusion.

In a solution with uniform concentration, the random diffusion of adsorbate in all directions leads to equal transfer rates into and out of any specific region. When a concentration gradient exists, more molecules are available in the high concentration region to diffuse towards the region of low concentration, and a net transfer of adsorbate occurs. It is the concentration difference between the solution phase which causes more adsorbate molecules to diffuse towards the adsorbent site and fewer to diffuse away, which leads to a mass transfer into and onto the adsorbent.

1.8.5 Pore Diffusion

Adsorbate molecules must diffuse through the static liquid in the pores of an adsorbent particle to reach the adsorption sites. The rate of diffusion through the solution in the macropores is usually assumed to be the same as the rate of diffusion through the static bulk solution since the adsorbate molecules collide with the pore walls infrequently relative to the rate of collision with solvent molecules. The rate of mass transfer through the pores varies inversely with the adsorbate concentration and inversely with the square of the particle radius. Mass transport becomes slower due to molecular size pore dimensions.

During adsorption of any adsorbates, the exterior surface of the particle becomes saturated with the adsorbent. A sharp boundary exists between saturated adsorbent and bare adsorbent. The adsorption zone becomes thicker as the time passes, with the boundary moving towards the centre of the particles. Adsorption and mass transfer terminate when the particles becomes fully saturated.

1.9   Adsorption Forces

During adsorption, adsorbates and adsorbent interact with each other. The nature of adsorption depends upon that force which acts between adsorbent and adsorbates. The adsorption forces decide the type of adsorption, whether it is physical or chemical. Adsorption force can be divided into van der Waals force and electrostatic force.

i. **Van der Waals Force:** The van der Waals force arises from rapidly fluctuating electron density in one molecule, which induces a complementary electrical moment in a nearby neighbour resulting in an alternative force between the two molecules. This force is responsible for the phenomenon of adsorption at interfacial surfaces. The potential energy ranges between 3–15 kcal/mole for readily adsorbable adsorbates. The degree of polarisability of the adsorbate in part determines the magnitude of the attractive potential; as a result, the potential experienced by different adsorbates approaching the same site on the surface is different. Readily polarisable molecules tend to be adsorbed most strongly. Also, the attractive potential for the same molecule may be different for different locations on the surface. The van der Waals force decreases very rapidly with the increase of distance away from the surface. As a result, an adsorbate adjacent to the surface experiences the full attractive force, while a second layer of adsorbate molecules on the top of the first will experience a much smaller attractive potential.

ii. **Electrostatic forces:** Some examples of adsorption where electrostatic forces act are ionic and polar adsorbates on ionic solids are water, amines, alcohols on inorganic salts, and oxides. Electrostatic forces are important and in some cases can even predominate. The adsorbate molecules arrange themselves on the surface with unlike charges nearest to each other, resulting in an electrostatic attraction which can be much stronger than the van der Walls attraction. If the adsorbate molecules do not have a permanent dipole, they can acquire an induced dipole when placed in an electrostatic field near an ionic solid surface. Unlike van der Waals forces, the electrostatic forces can be repulsive as well as attractive.

1.10   Adsorption Theories

Positive adsorption in a solid-liquid system results in the removal of solute from solution and their concentration at the surface of the solid, until such time as the concentration of the solute remaining in the solution is in a dynamic equilibrium with that at the surface.

At equilibrium, there is a defined distribution of solute between the liquid and solid phases. The distribution ratio is a measure of the position of equilibrium in the adsorption process; it may be a function of the nature of competing solutes, the nature of the solution, and so on. For depicting this distribution, express the quantity q_e, as a function of C at a fixed temperature, where q_e is the amount of solute adsorbed per unit weight of solid adsorbent, and C is the concentration of solute remaining in solution at equilibrium. This expression is termed an adsorption isotherm. The adsorption isotherm is a functional expression for the variation of adsorption with a concentration of adsorbate in bulk solution at constant temperature.

1.10.1   Classification of Adsorption Isotherms

The adsorption isotherms for dilute solutions are classified into four main categories according to the nature of the initial portion of the curve, and thereafter into subgroups, based on the shapes of the upper part of the curve [74].

The main classes are as follows:

    i. S-curve
   ii. L-curve (Langmuir type)
  iii. H-curve (high affinity)
  iv. C-curve (Constant partition)

The **S-class** of isotherm signifies an increase in affinity for adsorbate after the initial adsorption. The isotherm represents a situation in which (i) as the adsorption proceeds, it becomes easier for solute molecules to be adsorbed, and those already adsorbed on the surface at the most energetic sites assist further adsorption by intermolecular bonding within the adsorbed layer; (ii) the solvent molecules are strongly adsorbed, and (iii) the adsorbate is monofunctional, i.e., the molecular state of adsorbate has a single point of strong attachment in an aromatic ring system.

The **L-class** of isotherm is commonly known as the Langmuir type. In this type, as the adsorption proceeds, the most energetic surface sites are initially covered with adsorbate, and the case in which adsorption takes place decreases until the monolayer is complete.

The **H-class** of isotherm has a positive value on the adsorbent surface reflecting the high affinity for solute onto part of the adsorbent surface, and consequently the depletion of solute concentration in the bulk phase.

The **C-class** of isotherm indicates a constant partition of solute between solution and adsorbate. In some instances, after a critical concentration, the curve may show an abrupt change suggesting that sites are becoming available for adsorption.

Several types of isothermal adsorption relations are reported. Adsorption isotherms are important tools for the analysis of the adsorption process. The adsorption isotherm is the initial experimental setup to determine feasibility of adsorption treatment. By using a batch equilibrium test relating to the adsorbed per unit weight, the concentration of adsorbate remaining in the solution can be calculated. The Langmuir and Freundlich models have been widely used to describe the behaviour of the adsorbent-adsorbate couple. The BET isotherm is also reported for some of the systems.

2

Material and Methods

2.1 Adsorbent Collection and Storage

Sand is collected from the riverbed of the Son River (Bihar) of central India, which is the largest of the Ganges southern tributaries. The sand samples (Figure 2.1) were collected upstream from the shallow waters near the bank.

The samples were then stored in polyethylene bags, which were treated with 5% H_2SO_4 and 1M $KMNO_4$.

2.2 Adsorbent Modification

Silica sand was modified by soaking in 40% H_2SO_4 for a period of 4 hrs, and after that the treated sample of silica sand was washed repeatedly with deionized water. Simultaneously, the pH of the supernatant water for washing was checked, and the washing of treated silica sand was done until the pH of the supernatant water reached a neutral pH range. Finally,

Figure 2.1 Raw silica sand used in the experiments.

the treated sample of silica sand was dried in an oven until completely dry.

2.3   Preparation of Adsorbate Cr (VI) and Ni (II) Solution

A stock Cr (VI) solution of 1000 mg/L was prepared from potassium dichromate salt ($K_2Cr_2O_7$). The working solutions with different concentrations were prepared by appropriate dilutions of the stock solution. For the preparation of the stock solution of Ni (II), nickel sulphate ($NiSO_4.H_2O$) was used. The pH of the solution was adjusted to the required value by using 0.1 M NaOH or 0.1 M HCl.

2.4   Instrumentation

The instruments used during the experiments are as follows, along with a brief detail of their manufacturer:

Instrument	Model No. and Manufacturer
Water bath shaker	Narang Scientific Works Pvt. Ltd. (NSW-133)
Research centrifuge	REMI (R 24)

(Continued)

(Continued)

Instrument	Model No. and Manufacturer
Digital pH meter	IKON
Analytical single pan balance	K.Roy (K-15 SUPER)
Scanning electron microscope (SEM)	Jeol, Japan, JSM 6390LV
Atomic absorption spectrophotometer	Elico, Japan

2.5   Batch Adsorption Experiment

2.5.1   Adsorption Experiments

After development of the two nano-adsorbents, effects of various important parameters on the removal of selected metallic species by nano-adsorbents were studied. For the removal of selected metallic species, batch adsorption experiments were conducted. All the chemicals used in the present studies were of AR grade and were procured from Merck, Mumbai, India. The chemicals used in this study were ammonia solution, nickel sulphate, dimethylglyoxime, bromine water, hydrochloric acid, 95% ethyl alcohol, potassium dichromate, 1,5-diphenylcarbazide, sulphuric acid, phosphoric acid, nitric acid, sodium hydroxide, and sodium perchlorate.

Batch adsorption studies were conducted to determine the optimum conditions for the removal of selected metallic species, namely Ni (II) and Cr (VI), from aqueous solutions.

    a. A stock solution of Ni (II) was prepared by dissolving nickel sulphate in 1000 ml of distilled water. Then this solution was used to prepare a working solution of nickel Ni (II) of different concentrations.

b. A stock solution of chromium was prepared by dissolving potassium dichromate in 1000 ml distilled water, and used for preparation of chromium solution at various concentrations range.

The ionic strength of the aqueous solutions of metallic ions were maintained at 1.0×10^{-2} M $NaClO_4$. Batch experiments were conducted separately for both the selected metallic species, namely Cr(VI) and Ni(II), using both nano alumina and nano iron. Batch adsorption experiments were performed by adding different quantities of nano adsorbents with 50ml of aqueous solutions of metal solutions of varying concentration in 250 ml stoppard conical flasks.

All the adsorption experiments were conducted at 25°C (± 0.5), at a pH of working solution and agitation rate of 100 rpm on a shaking thermostat. After equilibrium time, the adsorbents were separated from the aqueous phase by centrifugation at 10,000 rpm for 10 minutes. The residual concentrations of Ni(II) in supernatants were determined by UV-visible spectrophotometer (Spectronic 20, Bausch & Lomb USA) at 445 nm, using the dimethylglyoxime (DMG) method by following the standard method for examination of water and wastewater [75]. While the residual concentration of Cr (VI) in aliquot was determined by using UV-visible spectrophotometer at 540 nm with 1,5 diphenylcarbazide.

2.5.2 Chromium (III) Analysis

For the determination of Cr (III) concentration, Cr (III) (formed due to the reduction of Cr (VI) into Cr (III) during the sorption process) was again converted to Cr (VI) by the addition of excess potassium permanganate at a high temperature (130–140°C) after the 1,5-diphenycarbazide was added. The pink coloured complex formed gives the concentration of Cr (VI) and Cr (III) which is total chromium. The Cr (III) concentration was then calculated by the difference of the total chromium and Cr (VI) concentrations measured above.

For this purpose, supernatant liquid after sorption was divided into two parts. In one sample, Cr (VI) concentration (C_e) was measured by the prescribed method. In contrast, another sample of Cr (VI) was determined after heating up the solution to 130–140°C with $KMnO_4$ solution. It was found that in both parts, the concentration of Cr (VI) was similar. This analysis confirms that the removal of Cr (VI) by nano adsorbents was only by the adsorption process. The reduction process is not involved in removal of Cr (VI) from aqueous solutions.

2.5.3 Analysis of Nickel

The residual concentrations of nickel in supernatants were determined by UV-visible spectrophotometer (Spectronic 20, Bosch & Lomb USA) at 445 nm using the dimethylglyoxime method [75].

The amount of metallic ions adsorbed per unit mass of the adsorbent was determined by using the following equation:

$$q = (C_i - C_e / W) * V \tag{2.1}$$

where q is the amount adsorbed on per unit mass of the adsorbent (mg/g), C_i, and C_e are the initial concentration and equilibrium concentration respectively (mg/L), W is the mass of adsorbent (g), and V (L) is the volume of solution.

The percentage removal of metallic ions was calculated by the following equation:

$$\% \text{ Removal of metallic ions} = (C_i - C_e / C_i) * 100 \tag{2.2}$$

After the development of nano adsorbents, batch experiments were conducted for removal of Cr (VI) and Ni (II) from aqueous solutions. Results of the effect of various important parameters were studied. By results of batch experiments, kinetic themodynamic and equilibrium studies were also carried out.

3

Results and Discussions

3.1 Characterization of Silica Sand

3.1.1 Chemical Characterization of Silica Sand

The chemical compositions are analyzed by EDS (Energy dispersive X-ray analyzer). The chemical characterization of silica sand shows silica to be its major constituent in the form of Silicon Oxide (SiO_2) as shown in Table 3.1. These oxides undergo surface hydroxylation (i.e., introduction of –OH group) in aqueous solution, which results in the formation of surface hydroxyl compounds. Their subsequent dissociation gives a negatively or positively charged surface as shown below in the mechanism reaction.

$$SiO_2 \longrightarrow >Si(OH)_2 \xrightarrow{H^+} Si^+\ OH \xrightarrow{H^+} Si^{+2}$$

Surface	**Cationic density**
Hydroxylation	**increases on**
	adsorbent surface

As pH of solution decreases, H^+ ions increases

$\longrightarrow$

45

Table 3.1 Chemical composition of silica sand (weight %)

Element	Raw Silica Sand	Modified Silica Sand
O	40.35	23.57
Al	0.77	7.56
Si	39.84	25.00
Au	19.04	11.98
S	0	0.06
K	0	13.38
N	0	18.45
Total	100	100

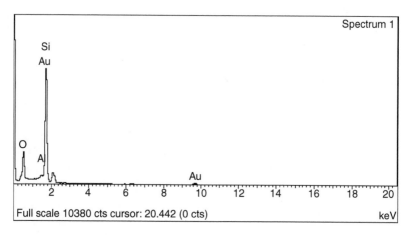

Figure 3.1 EDS spectrum of raw silica sand.

Following are figures (Figures 3.1 and 3.2) of X-ray analysis, which shows the peaks of respective elements found on the surface of silica sand.

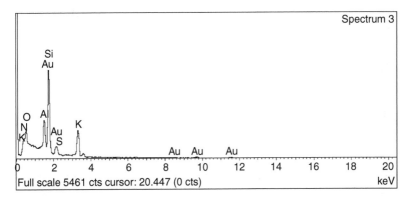

Figure 3.2 EDS spectrum of modified silica sand.

3.1.2 Fourier Transform Infrared Spectroscopy

The IR Spectra of raw and modified silica sand were measured on a Fourier Transform Infrared Spectrophotometer (Varian 1600 FT-IR Scimitar Series) to elucidate the functional group presenting on the surface of silica sand before and after activation. For measuring IR spectra, 5mg of raw silica sand and modified silica sand were encapsulated in 400mg of KBr. Translucent disk was made by pressing the ground mixed material with the hydraulic pallet press (PCI services, Bhandup, Mumbai) for 1 minute. The spectra were recorded in a FTIR within the range of 500–6000 cm^{-1} (Figure 3.3). The peaks found at 759.49 cm^{-1} shows the presence of silica (Si-O bond), and 1783 cm^{-1} shows the presence of 4-membered cyclic compound, which is probably of an organic in nature.

3.1.3 SEM Characterization of Silica Sand

The morphological characteristics of silica sand were evaluated using a scanning electron microscope (SEM). The scanning electron microscopy (SEM) observations were carried out with a Jeol, Japan, JSM-6390LV microscope equipped with an Oxford Link ISIS energy dispersive X-ray analyzer (EDS). An

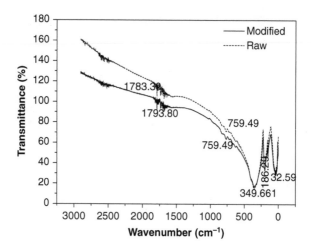

Figure 3.3 FTIR spectra of raw silica sand and modified silica sand before adsorption.

electron acceleration voltage of 20 kV was applied for SEM observation. The SEM measures the surface morphology of conducting and non-conducting materials by analyzing the backscattered electrons (BSE) and secondary electrons (SE). The chemical compositions are analyzed by EDS. Before analysis, the samples were coated with platinum in order to make them conductive.

Figure 3.4 shows the SEM micrographs of the raw silica sand at 100 times magnification, and unveils its rough surface. Figure 3.5 shows the SEM micrographs of the raw silica sand at 5000 times magnification. It can be seen from Figure 3.5 that the silica sand has a large number of porous structures because of fine crystals that are staggered in three-dimensional spaces. The surface is in a clean state which may ultimately increase the adsorption capacity of Cr (VI) on the silica sand.

Figures 3.6 and 3.7 show modified silica sand at ×100 and ×5000 magnifications respectively. These figures of modified silica sand show the roughness of the surface which is clearly visible in the ×5000 magnification as slopes and grooves.

Figure 3.4 SEM image of raw silica sand, showing rough surface at ×100 magnification.

Figure 3.5 SEM image of a single raw silica sand particle, showing reactive adsorption centre at ×5000 magnification.

These slopes and grooves increase the surface area, and thus the adsorption capacity of the adsorbent. Additionallly, the slopes and grooves act as a reactive adsorption centre for Cr and Ni ions adsorption.

Figure 3.6 SEM image of modified silica sand, showing rough surface at ×100 magnification.

Figure 3.7 SEM image of a modified silica sand particle, showing reactive adsorption centre at ×5000 magnification.

3.1.4   Determination of pH_{ZPC}

The pHzpc is an important property of adsorbent. This parameter determines the surface behaviour of adsorbent.

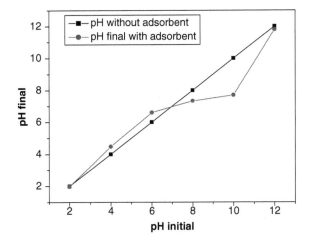

Figure 3.8 The pHzpc of raw silica sand (pHzpc of raw silica sand was found to be = 6.98).

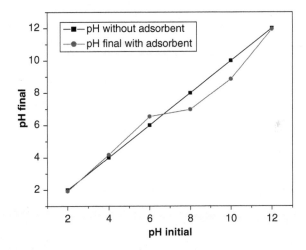

Figure 3.9 The pHzpc of modified silica sand (pHzpc of modified silica sand was found to be = 6.66).

The electrical neutrality of the adsorbent surface is indicated by pHzpc at a particular pH value.

In our study for the determination of pHzpc, we used 0.01 M NaCl solution, and the pH was adjusted in the range of

2–12 using NaOH and/or HCl solution. In each 50 ml of pH-adjusted 0.01 M NaCl sample, 0.20gm of silica sand (modified as well as raw) was added. After these samples were left for 48 hours, the final pH of the solutions was measured. Then graphs were plotted between pH final versus pH initial. The point of intersection of these two curves has been recorded as pHzpc of raw and modified silica sand. The graphs of the pHzpc of raw and modified silica sand are shown below in Figures 3.8 and 3.9 respectively.

3.2   Effect of Contact Time and Initial Concentration of Cr (VI) and Ni (II)

During the uptake of pollutants from wastewater by the adsorption process, the following two factors play important roles:

i.   initial concentration of the adsorbate in the solution, and
ii. time of contact between the adsorbate and adsorbent phases.

A rapid transport of adsorbate species from the bulk to the surface of the adsorbent cuts short the equilibrium period. The time taken to attain equilibrium is important in order to predict the efficiency and feasibility of an adsorbent for its use in water pollution control. The development of charge on the adsorbent surface is also governed by contact time. The knowledge of the rate of uptake of molecules and ions to attain thermodynamic equilibrium is necessary for complete characterization of the adsorbent.

The effect of contact time was performed on an initial concentration of 5mg/l, 10mg/l, and 15mg/L at pH 2.5 for Cr (VI) and pH 6.0 for Ni (II) by using 1g of modified silica sand for the contact time periods ranging from 10–120 min at 25°C. After determining the equilibrium contact time (i.e., the period after which no adsorption takes place), all the further experiments are limited up to that time.

Table 3.2 Effect of initial concentration on the removal of Cr (VI) by adsorption on modified silica sand (pH = 2.5, temp = 25°C, dose = 1gm/50ml adsorbent, rpm = 150).

Contact Time (min)	Initial Concentration (mg L^{-1})					
	5		10		15	
	Amount Adsorbed (mgg^{-1})	Removal (%)	Amount Adsorbed (mgg^{-1})	Removal (%)	Amount Adsorbed (mgg^{-1})	Removal (%)
20	0.077	30.88	0.102	20.40	0.065	8.75
40	0.129	51.42	0.240	48.00	0.251	33.48
60	0.173	69.12	0.340	68.00	0.488	65.16
80	0.201	80.40	0.347	69.406	0.504	67.24
100	0.201	80.40	0.347	69.406	0.504	67.24

During the removal of Cr (VI) by adsorption on silica sand, the uptake increased up to 80 min for Cr (VI), and after this the rate of uptake become constant (Table 3.2; Figure 3.10). It is also clear from the figure that by decreasing the concentration of the solution from 15 mg/L to 5 mg/L, the removal (%) increased from 67.24% to 80.40% at pH 2.5 at 25°C. It can also be observed that for the present system, the equilibrium is independent of the adsorbate concentration. The curves obtained are single and smooth indicating the formation of monolayer on the surface of adsorbent during the removal process. In the initial stages the slope of the plots is greater, and it decreases with time. This reveals that the rate of uptake is rapid in early stages, and gradually decreases and becomes constant when the equilibrium is attained. Furthermore, the higher (%) removal in low concentration ranges is of industrial importance.

In the case of Ni (II) removal by modified silica sand, the removal increased up to 120 min, after that the rate of uptake becomes constant (Table 3.3; Figure 3.11) . It is also clear from the figure that by decreasing the concentration of the

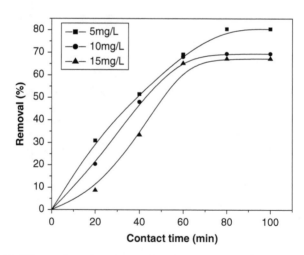

Figure 3.10 Effects of contact time and concentration on the removal (%) of Cr (VI) by adsorption on modified silica sand (pH = 2.5, temp = 25°C, dose = 1gm/50ml adsorbent, rpm = 150).

Table 3.3 Effect of initial concentration on the removal of Ni (II) by adsorption on modified silica sand (pH = 6.0, temp = 25°C, dose = 1gm/50ml, rpm = 150).

| Contact Time (min) | Initial Concentration (mg L⁻¹) | | | | | |
| | 5 | | 10 | | 15 | |
	Amount Adsorbed (mgg⁻¹)	Removal (%)	Amount Adsorbed (mgg⁻¹)	Removal (%)	Amount Adsorbed (mgg⁻¹)	Removal (%)
30	0.067	26.84	0.099	19.84	0.093	12.34
60	0.107	42.76	0.181	36.29	0.231	30.76
90	0.166	66.76	0.306	61.27	0.401	53.46
120	0.172	68.76	0.311	62.24	0.406	54.09
150	0.172	68.76	0.311	62.24	0.406	54.09

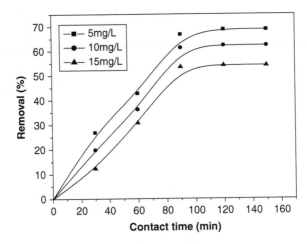

Figure 3.11 Effects of contact time and concentration on the removal (%) of Ni (II) by adsorption on modified silica sand (pH = 6.0, temp = 25°C, dose = 1gm/50ml adsorbent, rpm = 150).

solution from 15 mg/L to 5 mg/L, the removal (%) increased from 54.09% to 68.76% at pH6.0, at 25°C and dose of 1 gm modified silica sand in 50ml of Ni (II) solution. The curves obtained are single and smooth, indicating the formation of monolayer on the surface of adsorbent during the removal process.

3.3   Effect of pH on the Removal of Cr (VI) and Ni (II)

The pH of solutions plays a very important role in the removal of pollutants from water and wastewater in a system using oxides, aluminosilicates, clays, and sand as adsorbents. It controls the surface charge and surface properties of the adsorbent surface as well as the degree of ionization. The effect of initial pH was investigated by agitating 50 ml Cr (VI) solutions having 5mg/L concentration at different pH values 2.5, 4.0, and 8.0 for Cr (VI), and pH 4.0, 6.0, and

8.0 for Ni (II) with 1.0 g silica sand in a 150 ml reagent bottle held on a mechanical shaker equipped with a thermostatic water bath for 120 min at 25°C. The initial pH of the solutions was adjusted with 0.1 M NaOH or 0.1 M HCl solutions using a pH meter. The reagent bottles were then shaken at 150 rpm. The initial concentration of solutions for Cr (VI) and Ni (II) were 5mg/L. Other experimental conditions were the same as in other cases.

In the case of removal of Cr (VI), the maximum removal is 80.40, 44.08, and 36.40% at pH values of 2.5, 4.0, and 8.0 respectively (Table 3.4; Figure 3.12). The curves are single and smooth and indicate the fitness of the adsorbent for the system. The figure also shows that in an acidic region, the removal of Cr (VI) by silica sand is maximum at pH 2.5.

The maximum removal Cr (VI) around pH 2.5 may be associated with the dissolution of the substrate and its interactions with $HCrO_4^-$ species leading to surface compound formation:

$$M + HCro_4^- \Leftrightarrow [M^{z-1}HCro_4]$$

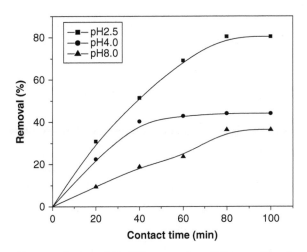

Figure 3.12 Time variation of Cr (VI) removal at different pH values by modified silica sand (Cr(VI) concentration = 5mg/L, temp = 25°C, dose = 1gm/50ml adsorbent, rpm = 150).

Table 3.4 Effect of pH on the removal of Cr (VI) by adsorption on modified silica sand (conc^n = 5mg/L, T = 25°C, dose = 1gm/50ml, rpm = 150).

Contact Time (min)	pH 2.5		pH 4.0		pH 8.0	
	Amount Adsorbed (mgg^{-1})	Removal (%)	Amount Adsorbed (mgg^{-1})	Removal (%)	Amount Adsorbed (mgg^{-1})	Removal (%)
20	0.077	30.88	0.056	22.36	0.023	9.38
40	0.129	51.42	0.101	40.32	0.047	18.84
60	0.173	69.12	0.107	42.88	0.059	23.58
80	0.201	80.40	0.110	44.08	0.091	36.40
100	0.201	80.40	0.110	44.08	0.091	36.40

Table 3.5 Effect of pH on the removal of Ni (II) by adsorption on modified silica sand (conc^n = 5 mg/L, temp. = 25°C, dose = 1gm/50ml, rpm = 150).

Contact Time (min)	pH 4.0		pH 6.0		pH 8.0	
	Amount Adsorbed (mgg^{-1})	Removal (%)	Amount Adsorbed (mgg^{-1})	Removal (%)	Amount Adsorbed (mgg^{-1})	Removal (%)
30	0.014	5.82	0.067	26.84	0.006	2.28
60	0.046	18.34	0.107	42.76	0.024	9.4
90	0.059	23.98	0.166	66.76	0.051	20.28
120	0.062	24.70	0.172	68.76	0.058	23.18
150	0.062	24.70	0.172	68.76	0.058	23.18

where M stands for the substrate. It is argued that at pH 2.5, a significantly high electrostatic attraction exists between adsorbent and adsorbate resulting in the formation of surface compound. Thus, it seems that coulombic attraction as well as surface complexation have important roles during the removal of Cr (VI) from water.

In the case of removal of Ni (II), the maximum removal was found to be 24.70, 68.76 and 23.18% at pH values of 4.0, 6.0, and 8.0 respectively (Table 3.5; Figure 3.13). The curves obtained are single and smooth and indicate the fitness of the adsorbent for the system. The figure also shows that in slightly acidic regions, the removal of Ni (II) by modified silica sand is maximum at pH 6.0.

3.4   Effect of Temperature on the Removal of Cr (VI) and Ni (II)

Temperature has an important role on the process of adsorption, and thus the removal of metal ions from the aqueous

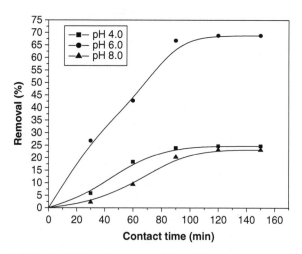

Figure 3.13 Effects of pH on the removal (%) of Ni (II) by adsorption on modified silica sand (conc^n = 5 mg/L, temp = 25°C, dose = 1gm/50ml adsorbent, rpm = 150).

solutions and wastewaters. Mostly adsorption is an exo-thermic process, but some examples of endothermic adsorp-tion have also been reported. The study undertaken here is governed by exothermic adsorption (i.e., lower temperature favours).

For isothermal studies, a series of reagent bottles containing 50 ml Cr (VI) solutions in the range of 5mg/L, 10mg/L, and 15mg/L were prepared. The weighed amount of 1g of silica sand was added to each reagent bottle and the mixtures were agitated at a constant temperature of 25°C, 30°C, and 35°C. These experiments were carried out at a constant pH of 2.5 for Cr (VI), and at a pH of 6.0 for Ni (II), for a contact period of 120 min.

At the end of the required contact period, the aqueous phase was separated from silica sand by centrifugation at 10,000 rpm for 10 min. Supernatants were then analyzed by an atomic absorption spectrophotometer.

In the case of Cr (VI) the removal decreased from 80.40% to 62.42% with an increase in temperature from 25°C to 35°C at 5mg/L Cr (VI) concentration, 69.406% to 40.20% with increase in temperature from 25°C to 35°C at 10mg/L Cr (VI) concentration, and 67.24% to 54.28% with an increase in tem-perature from 25°C to 35°C at 15mg/L Cr (VI) concentration as shown in Tables 3.6, 3.7 and 3.8, and Figures 3.14, 3.15, 3.16 respectively.

From the study of these (%) removal graphs, it can be seen that the maximum adsorption of Cr (VI) takes place at the lowest initial Cr (VI) concentration (i.e., 5 mg/L) at 25°C.

The decrease in removal at increasing values of temperature can be attributed to the fact that at an enhanced temperature, the relative escaping tendency of Cr (VI) species gets enhanced, which consequently results in the reduction of boundary layer thickness. Also, solubility of Cr (VI) increases with increasing temperatures, which is also a reason of the decrease in removal at an increasing value of temperature.

Table 3.6 Effect of temperature on the removal of Cr (VI) by adsorption on modified silica sand (conc^n = 5mg/L, pH = 2.5, dose =1gm/50ml adsorbent, rpm = 150).

Contact Time (min)	Initial Concentration 5mg/L					
	Temperature (°C)					
	25		30		35	
	Amount Adsorbed (mgg^{-1})	Removal (%)	Amount Adsorbed (mgg^{-1})	Removal (%)	Amount Adsorbed (mgg^{-1})	Removal (%)
20	0.077	30.88	0.041	16.22	0.019	7.84
40	0.129	51.42	0.100	40.04	0.048	19.38
60	0.173	69.12	0.159	63.56	0.095	38.16
80	0.201	80.40	0.196	78.52	0.156	62.42
100	0.201	80.40	0.196	78.52	0.156	62.42

Table 3.7 Effect of temperature on the removal of Cr (VI) by adsorption on modified silica sand (conc^n = 10mg/L, pH = 2.5, dose = 1gm/50ml adsorbent, rpm = 150).

Contact Time (min)	Initial Concentration 10mg/L					
	Temperature (°C)					
	25		30		35	
	Amount Adsorbed (mgg^{-1})	Removal (%)	Amount Adsorbed (mgg^{-1})	Removal (%)	Amount Adsorbed (mgg^{-1})	Removal (%)
20	0.102	20.40	0.062	12.32	0.046	9.25
40	0.240	48.00	0.164	32.73	0.093	18.57
60	0.340	68.00	0.293	58.72	0.200	40.13
80	0.347	69.406	0.294	58.80	0.201	40.20
100	0.347	69.406	0.294	58.80	0.201	40.20

Table 3.8. Effect of temperature on the removal of Cr (VI) by adsorption on modified silica sand ($conc^n$ = 15 mg/L, pH = 2.5, dose = 1gm/50ml adsorbent, rpm = 150)

Contact Time (min)	Initial Concentration 10mg/L					
	Temperature (°C)					
	25		30		35	
	Amount Adsorbed (mgg^{-1})	Removal (%)	Amount Adsorbed (mgg^{-1})	Removal (%)	Amount Adsorbed (mgg^{-1})	Removal (%)
20	0.065	8.75	0.056	7.49	0.044	5.866
40	0.251	33.48	0.221	29.513	0.183	24.40
60	0.488	65.16	0.454	60.66	0.406	54.16
80	0.504	67.24	0.455	60.686	0.407	54.28
100	0.504	67.24	0.455	60.686	0.407	54.28

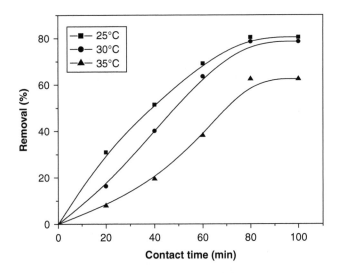

Figure 3.14 Time variation of Cr (VI) removal at different temperature values by modified silica sand (conc^n = 5 mg/L, pH = 2.5, dose = 1gm/50ml adsorbent, rpm = 150).

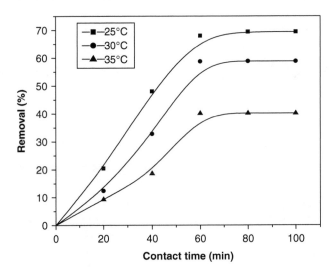

Figure 3.15 Time variation of Cr (VI) removal at different temperature values by modified silica sand (conc^n =10 mg/L, pH = 2.5, dose =1gm/50ml adsorbent, rpm = 150).

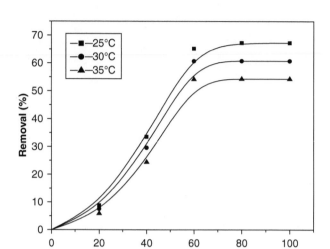

Figure 3.16 Time variation of Cr (VI) removal at different temperature values by modified silica sand (conc^n = 15mg/L, pH =2.5, dose = 1gm/50ml adsorbent, rpm = 150).

In the case of Ni (II), the removal decreased from 68.76% to 39.48% with increase in temperature from 25°C to 35°C at 5mg/L Ni (II) concentration, 62.24% to 31.21% with increase in temperature from 25°C to 35°C at 10mg/L Ni (II) concentration, and 54.09% to 49.27% with increase in temperature from 25°C to 35°C at 15mg/L Ni (II) concentration as shown in Figures 3.17, 3.18, and 3.19 respectively. The data is also presented in Tables 3.9, 3.10 and 3.11 respectively.

3.5   Effect of Adsorbent Dosage on the Removal of Cr (VI) and Ni (II)

In order to determine the effect of a dose three samples were taken, each having 50 ml of 5mg/L Cr (VI) solution, and a dose of modified silica sand that varied from 1gm/50ml, 1.5mg/50ml, to 2.0 mg/50ml. The samples were agitated in a bath shaker for 120 min at a constant temperature of 25°C.

Table 3.9 Effect of temperature on the removal of Ni (II) by adsorption on modified silica sand (conc^n = 5mg/L, pH = 6.0, dose = 1gm/50ml adsorbent, rpm = 150).

Contact Time (min)	Initial Concentration 5mg/L					
	Temperature (°C)					
	25		30		35	
	Amount Adsorbed (mgg^{-1})	Removal (%)	Amount Adsorbed (mgg^{-1})	Removal (%)	Amount Adsorbed (mgg^{-1})	Removal (%)
30	0.067	26.84	0.044	17.62	0.034	13.50
60	0.107	42.76	0.085	34.06	0.057	22.96
90	0.166	66.76	0.156	62.42	0.095	37.84
120	0.172	68.76	0.157	62.56	0.099	39.48
150	0.172	68.76	0.157	62.56	0.099	39.48

Table 3.10 Effect of temperature on the removal of Ni (II) by adsorption on modified silica sand (conc^n =10 mg/L, pH = 6.0, dose = 1gm/50ml adsorbent, rpm = 150).

Contact Time (min)	Initial Concentration 10mg/L					
	Temperature (°C)					
	25		30		35	
	Amount Adsorbed (mgg^{-1})	Removal (%)	Amount Adsorbed (mgg^{-1})	Removal (%)	Amount Adsorbed (mgg^{-1})	Removal (%)
30	0.099	19.84	0.056	11.19	0.049	9.96
60	0.181	36.29	0.149	29.98	0.120	24.02
90	0.306	61.27	0.274	54.88	0.155	31.08
120	0.311	62.24	0.278	55.51	0.156	31.21
150	0.311	62.24	0.278	55.51	0.156	31.21

Table 3.11 Effect of temperature on the removal of Ni (II) by adsorption on modified silica sand (conc^n = 15mg/L, pH = 6.0, dose=1gm/50ml adsorbent, rpm = 150).

Contact Time (min)	Initial Concentration 15mg/L					
	Temperature (°C)					
	25		30		35	
	Amount Adsorbed (mgg^{-1})	Removal (%)	Amount Adsorbed (mgg^{-1})	Removal (%)	Amount Adsorbed (mgg^{-1})	Removal (%)
30	0.093	12.34	0.075	10.02	0.071	9.51
60	0.231	30.76	0.201	26.78	0.184	24.52
90	0.401	53.46	0.394	52.63	0.356	47.50
120	0.406	54.09	0.394	52.70	0.369	49.27
150	0.406	54.09	0.394	52.70	0.369	49.27

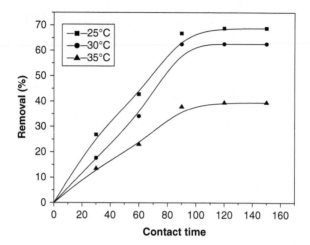

Figure 3.17 Effect of temperature on the removal (%) of Ni (II) by adsorption on modified silica sand (conc^n = 5mg/L, pH = 6.0, dose = 1gm/50ml adsorbent, rpm = 150).

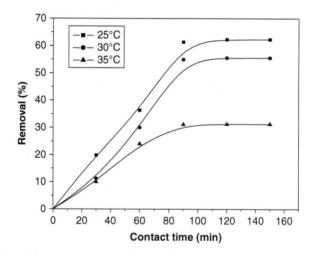

Figure 3.18 Effect of temperature on the removal (%) of Ni (II) by adsorption on modified silica sand (conc^n = 10 mg/L, pH = 6.0, dose = 1gm/50ml adsorbent, rpm = 150).

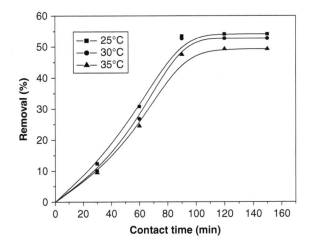

Figure 3.19 Effect of temperature on the removal (%) of Ni (II) by adsorption on modified silica sand (conc^n =15 mg/L, pH = 6.0, dose = 1gm/50ml adsorbent, rpm = 150).

The effect of the modified silica sand dosage is depicted in Tables 3.12 and 3.13 and Figures 3.20 and 3.21. Removal efficiency was found to increase proportionally with the amount of the modified silica sand dosage until a certain value was reached; afterwards, the removal efficiency remained constant even if the silica sand is added. It is apparent that the percent removal of heavy metals increases rapidly with an increase in dose of the silica sand due to the greater availability of the adsorption sites on the surface area. The decrease in adsorption uptake with an increasing adsorbent dosage is mainly due to unsaturation of adsorption sites through the adsorption reaction. At 1gm/50ml of modified silica sand dosage level, the removal of Cr (VI) was found to be 80.40%. Removal efficiency of 81.08% was achieved at the highest adsorbent dosage of 2gm/50ml (Figure 3.20).

In the case of Ni (II) removal at a 1gm/50ml dose, the removal percentage was found to be 68.76%, at a 1.5 gm/50ml dose removal the percentage was 78.72%, but with a further increase in dose, i.e., 2gm/50ml, the removal percentage decreases to 78.04%. The results are shown in Figure 3.21.

Table 3.12 Effect of dose on the removal of Cr (VI) by adsorption on modified silica sand (conc^n = 5 mg/L, pH = 2.5, temp = 25°C, rpm = 150).

Contact Time (min)	Adsorbent Dose (g)					
	1.0		1.5		2.0	
	Amount Adsorbed (mgg^{-1})	Removal (%)	Amount Adsorbed (mgg^{-1})	Removal (%)	Amount Adsorbed (mgg^{-1})	Removal (%)
20	0.077	30.88	0.061	36.49	0.053	42.38
40	0.129	51.42	0.089	53.513	0.073	58.2
60	0.173	69.12	0.119	71.66	0.098	78.26
80	0.201	80.4	0.135	80.886	0.100	80.08
100	0.201	80.4	0.135	80.886	0.100	80.08

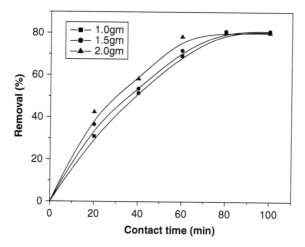

Figure 3.20 Time variation of Cr (VI) removal at different dosage of modified silica sand (conc^n = 5mg/L, pH = 2.5, temp = 25°C, rpm = 150).

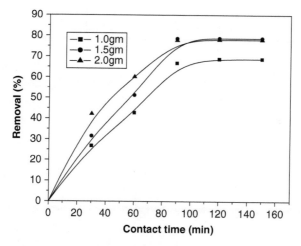

Figure 3.21 Effect of dose on the removal (%) of Ni (II) by adsorption on modified silica sand (conc^n = 5mg/L, pH = 6.0, temp = 25°C, rpm = 150).

3.6   Adsorption Isotherm

The adsorption isotherm is an equation relating the amount of solute adsorbed onto the solid and the equilibrium

Table 3.13 Effect of dose on the removal of Ni (II) by adsorption on modified silica sand ($conc^n$ = 5mg/L, pH = 6.0, temp = 25°C, rpm = 150).

Contact Time (min)	Adsorbent Dose (g)					
	1.0		1.5		2.0	
	Amount Adsorbed (mgg^{-1})	Removal (%)	Amount Adsorbed (mgg^{-1})	Removal (%)	Amount Adsorbed (mgg^{-1})	Removal (%)
30	0.067	26.84	0.052	31.54	0.052	42.22
60	0.107	42.76	0.085	51.26	0.073	60.22
90	0.166	66.76	0.131	78.60	0.097	77.82
120	0.172	68.76	0.131	78.72	0.098	78.04
150	0.172	68.76	0.131	78.72	0.098	78.04

concentration of the solute in a solution at a given temperature. The adsorption capacity of adsorbent is determined by the use of an adsorption isotherm.

The most commonly used isotherms for the application of adsorbent in wastewater treatment are the Freundlich and Langmuir isotherms.

3.6.1 Langmuir Isotherm Model

According to the Langmuir model, adsorption occurs uniformly on the active sites of the adsorbent, and once an adsorbate occupies a site no further adsorption can take place at this site.

The Langmuir model is given by the following equation [76, 77]:

$$\frac{C_e}{q_e} = \frac{1}{q_{max}K_L} + \frac{C_e}{q_{max}} \qquad (3.1)$$

where, q_e is the concentration of adsorbate in the solid phase at equilibrium (mg/L), C_e is the concentration of adsorbate in the liquid phase at equilibrium (mg/L), and q_{max} (mg/g) and K_L (L/mg) are the Langmuir constants. The slope and the intercept of this line give the values of q_{max} and K_L.

The essential characteristics of the Langmuir equation can be expressed in terms of a dimensionless separation factor, R_L, defined as:

$$R_L = \frac{1}{1 + K_L C_0} \qquad (3.2)$$

where, C_0 is the highest initial solute concentration and K_L is the Langmuir's adsorption constant (L/mg).The parameter R_L indicates the shape of isotherm in the following manner [78, 79]:

3.6.2 Assumptions in Langmuir Isotherm Model

Langmuir had to make several assumptions in order to derive an (admittedly over-simplified) adsorption expression:

Value of R_L	Type of Isotherm
$R_L > 1$	Unfavourable
$R_L = 1$	Linear
$0 < R_L < 1$	Favourable
$R_L = 0$	Irreversible

- All surface sites have the same adsorption energy for the adsorbate (and a similar statement for the solvent).
- Adsorption (of either solvent or adsorbate) at one site does not affect the availability of (block) the next site to adsorb solvent or adsorbate.
- Adsorption (of either solvent or adsorbate) at one site does not affect the energy of adsorption of the neighboring sites (as they adsorb either solvent or adsorbate).
- The activity of the adsorbate is directly proportional to its concentration (and a similar statement for the solvent) [61].

These assumptions are not satisfied for many real life cases. However, the simplified version gives insight into the dependence of surface coverage on concentration in solution.

Table 3.14 shows the values of R_L [0.18–0.24 for Cr (VI)] and [0.23–0.37 for Ni (II)], which were in range of 0–1 at all temperatures studied, which confirms the favorable uptake of Cr (VI) (Figure 3.22) and Ni (II) (Figure 3.23). R^2 values further reveal that Ni (II) adsorption favors the Langmuir isotherm while Cr (VI) does not favor the Langmuir isotherm.

3.6.3 Freundlich Isotherm Model

The Freundlich isotherm is an empirical model that is based on adsorption on a heterogeneous surface. This is

Table 3.14 Values of langmuir constants for adsorption of chromium (VI) and nickel (II) on modified silica sand at different temperatures.

Adsorbate	Temperature (°C)	q_{max} (mg/g)	K_L (l/mg)	R^2	R_L
Cr (VI)	25	0.79	0.31	0.91	0.18
	30	0.59	0.36	0.87	0.15
	35	0.49	0.21	0.68	0.24
Ni (II)	25	0.67	0.22	0.99	0.23
	30	0.86	0.11	0.96	0.37
	35	0.24	0.23	0.95	0.22

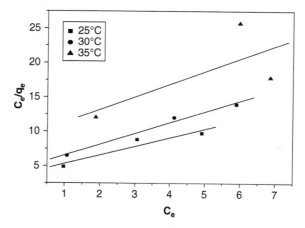

Figure 3.22 Langmuir isotherm for adsorption of chromium (VI) on modified silica sand at different temperatures.

applicable to a non-ideal sorption as well as a multilayer sorption process [80].

The Freundlich model is given by the following equation:

$$\log q_e = \log K_f + \frac{1}{n}\log C_e \tag{3.3}$$

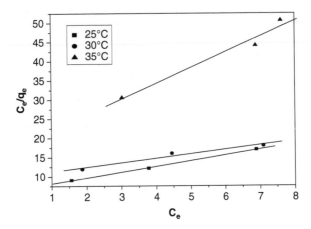

Figure 3.23 Langmuir isotherm for adsorption of nickel (II) on modified silica sand at different temperatures.

where, q_e is the concentration of adsorbate in solid phase at equilibrium (mg/L), C_e is the concentration of adsorbate in liquid phase at equilibrium (mg/L), K_f [(mg/g)(L/mg)$^{1/n}$] is the Freundlich constant, and $1/n$ (dimensionless) is the heterogeneity factor.

The Freundlich constants were determined from the slope and intercept of a plot of log q_e versus log C_e [81]. Values, n>1, represents favourable adsorption condition [82].

Adsorption studies were carried out on two isotherm models: the Langmuir and Freundlich isotherm models. The applicability of the isotherm model is determined by comparing the correlation coefficients, R^2, values. The isotherm model from which the values obtained are nearer to 1.0 is the best fitted model to the given sorption process.

Table 3.15 shows the values of Freundlich constants for chromium and nickel. The table shows that exponents n were greater than 1, and n>1 represent favorable adsorption condition. R^2 values further reveals that Cr (VI) and Ni (II) adsorption favours the Freundlich isotherm. Figures 3.24 and 3.25 show the Freundlich isotherm for the adsorption of Cr (VI) and Ni (II), respectively.

Table 3.15 Values of freundlich constants for adsorption of chromium (VI) and Nickel (II) on modified silica sand at different temperatures.

Adsorbate	Temperature (°C)	KF (mg/g) (L/mg)$^{1/n}$	n	R^2
Cr (VI)	25	0.19	1.75	0.98
	30	0.18	2.26	0.88
	35	0.11	1.86	0.89
Ni (II)	25	0.13	1.71	0.98
	30	0.10	1.45	0.99
	35	0.03	1.02	0.88

3.7 Adsorption Kinetics

The study of adsorption kinetics is important because it provides valuable information and insights into the reaction pathways and the mechanism of the reactions. An adsorption process is normally controlled by: (i) Transport of the solute from solution to the film surrounding the adsorbent, (ii) from the film to the adsorbent surface, and (iii) from the surface to the internal sites followed by binding of the metal (ions) to the active sites. The slowest steps out of all the steps determine the overall rate of the adsorption process. Usually in most adsorption processes, (ii) accounts for the surface adsorption, and (iii) leads to intraparticle adsorption [83]. To investigate the mechanism of chromium and nickel adsorption on modified silica sand, the pseudo-first-order, pseudo-second-order, and intraparticle diffusion equations were used.

3.7.1 Pseudo-first-order Kinetic Model

A simple kinetic analysis of adsorption is the pseudo-first-order equation in the form [84, 85]:

$$\frac{dq_t}{dt} = k_1 (qe - qt) \qquad (3.4)$$

where, k_1 is the rate constant of pseudo-first-order adsorption, and qe represents adsorption capacity (i.e., the amount of

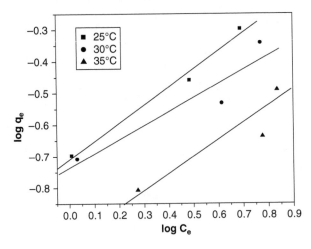

Figure 3.24 Freundlich isotherm for adsorption of chromium (VI) on modified silica sand at different temperatures.

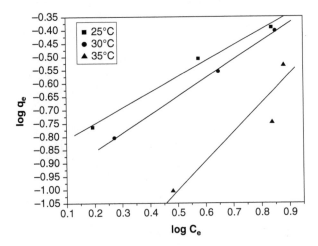

Figure 3.25 Freundlich isotherm for adsorption of nickel on modified silica sand at different temperatures.

adsorption corresponding to monolayer coverage). After defi-
nite integration by applying the initial conditions $t = 0$ to t and
$q_t = 0$ to q_t, eq.(4) [85] becomes:

$$\log(q_e - q_t) = \log q_e - \frac{k_1}{2.303}t \qquad (3.5)$$

where, q_e and q_t are the amount of heavy metal adsorbed
(mg/g) at equilibrium, and at any time t, k_1 is the rate constant
(min^{-1}). The values of the pseudo-first-order rate constant
k_1 were obtained from the slopes of linear plot of log(q_e–q_t)
versus t.

The k_1 values, the correlation coefficients, and R^2 values are
given in Table 3.16 for chromium as well as nickel, and corresponding graphs are shown in Figures 3.26 and 3.27. The values
of R^2 shows that Cr (VI) and Ni (II) adsorption favours pseudo-
first-order kinetics.

3.7.2   Pseudo-second-order Kinetic Model

The pseudo-second-order kinetic model can be represented in
the following form:

$$\frac{dq_t}{d_t} = k_2(q_e^2 - q_t)^2 \qquad (3.6)$$

Table 3.16 Pseudo-first-order constants for adsorption of
chromium (VI) and nickel (II) on modified silica sand at different
temperatures.

Adsorbate	Temperature (°C)	k_1 (min^{-1})	R^2
Cr (VI)	25	0.0368	0.97
	30	0.0358	0.96
	35	0.0203	0.95
Ni (II)	25	0.048	0.87
	30	0.079	0.82
	35	0.046	0.86

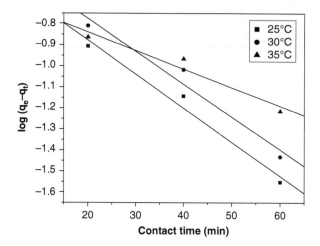

Figure 3.26 Pseudo-first-order kinetic plots for the adsorption of chromium (VI) on modified silica sand at different temperatures.

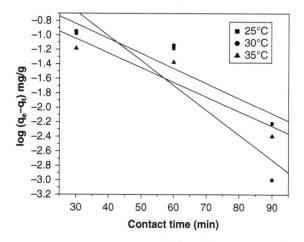

Figure 3.27 Pseudo-first-order kinetic plots for the adsorption of nickel (II) on modified silica sand at different temperatures.

where k_2 is the rate constant of pseudo-second-order adsorption. After the integration of eq. (6) [86], and applying the initial conditions we have:

$$\frac{t}{q_t} = \frac{1}{k_2 q e^2} + \frac{1}{q_e} t \qquad (3.7)$$

where k_2 (min g/mg) is the rate constant for the pseudo-second-order adsorption kinetics. The slopes of the plots t/q_t versus t give the value of q_e, and from the intercept k_2 can be calculated.

The pseudo-second-order rate constants k_2 and the corresponding linear regression correlation coefficients values R^2 are given in Table 3.17. Corresponding graphs are shown in Figures 3.28 and 3.29.

Table 3.17 Values of rate constant of pseudo-second-order rate for the removal of chromium (VI) and nickel (II) by adsorption on modified silica sand.

Adsorbate	Temperature (°C)	k_2 (g/mg min)	R^2
Cr (VI)	25	0.0236	0.97
	30	0.0035	0.49
	35	0.0684	0.97
Ni (II)	25	0.006	0.61
	30	0.004	0.58
	35	0.001	0.14

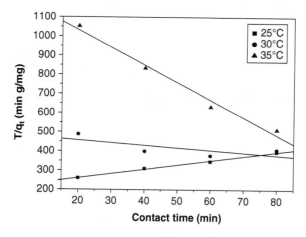

Figure 3.28 Pseudo-second-order kinetic plots for the adsorption of chromium on modified silica sand at different temperatures.

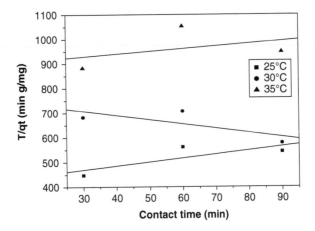

Figure 3.29 Pseudo-second-order kinetic plots for the adsorption of nickel on modified silica sand at different temperatures.

The values for linear regression coefficients R^2 were obtained, and for the pseudo-first-order kinetic model it was found to be $R^2 \geq 0.95$ for chromium and $R^2 \geq 0.82$ for nickel. This suggests that the adsorption kinetics of removal of Ni (II) and Cr (VI) follows the pseudo-first-order kinetic model.

3.7.3 Intraparticle Diffusion Model

In adsorption systems, when there is the possibility of intraparticle diffusion being the rate-limiting step, the intraparticle diffusion approach is described by many scientific workers. The rate constant for intraparticle diffusion (k_{dif}) is determined using the following equation [87]:

$$q_t = k_{dif} \sqrt{t} + C \qquad (3.8)$$

where k_{dif} is the intraparticular diffusion rate constant (mg/g min$^{1/2}$). If the intraparticle diffusion is rate-limited, then plots of adsorbate uptake q_t versus the square root of time ($t^{1/2}$) would result in a linear relationship, and k_{dif} and C values can be obtained from these plots (Figures 3.30 and 3.31) and are shown in Table 3.18. The first step in the diffusion model is the mass transfer of adsorbates molecules from bulk solution to the adsorbent surface, and the second stage is the intraparticle diffusion on silica sand. Moreover, the particle diffusion would be the rate controlling step if the lines pass through the origin, but here it is not the rate controlling step as concluded by Figures 3.30 and 3.31.

Table 3.18 Intraparticle diffusion constants for different temperatures at initial chromium and nickel concentration of 5 mg/l.

Adsorbate	Temperature(°C)	kdif (mg/g min$^{1/2}$)	R^2
Cr (VI)	25°C	0.028	0.99
	30°C	0.035	0.99
	35°C	0.030	0.94
Ni (II)	25°C	0.021	0.95
	30°C	0.023	0.93
	35°C	0.013	0.95

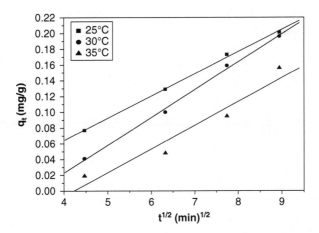

Figure 3.30 Intraparticle diffusion kinetics for adsorption of chromium (VI) on modified silica sand at different temperatures (initial chromium concentration of 5mg/L).

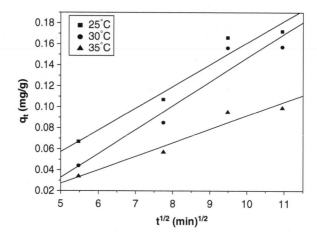

Figure 3.31 Intraparticle diffusion kinetics for adsorption of nickel (II) on modified silica sand at different temperatures (initial nickel concentration of 5mg/L).

3.8   Thermodynamic Studies

Thermodynamic studies for the processes are important because they provide vital information regarding various aspects. As far as adsorption studies are concerned, scientific workers have mainly concentrated on thermodynamic studies by calculating values of various thermodynamic parameters [88–91] and describing the processes. Thermodynamic parameters such as changes in standard free energy (ΔG^0), standard enthalpy (ΔH^0), and standard entropy (ΔS^0) are studied to determine the spontaneity (practical feasibility) of a given adsorption process. Negative values of free energy indicate spontaneity of the process. In some cases the values of ΔG^0 have been reported to be positive. The authors [92] emphasize that the values of free energy calculated from the uncontrolled pH experiments data were negative, whereas they were positive for the experiments carried out under controlled optimum conditions. The authors further claim that this indicates the importance of the conditions of the adsorption medium besides the physical and chemical properties of adsorbate and adsorbent. Others [93–95] have also reported on thermodynamic studies for the removal of metallic species. Thermodynamic and kinetic studies of the removal of nickel by cashew nut shell [96], clinoptilolite [97], nanostructured hydrous titanium(IV) oxide [98], and multi-walled carbon nanotubes [99] have all been reported on.

The pseudo-second-order rate constant of heavy metal adsorption is expressed as a function of temperature of Arrhenius-type relationship [99]:

$$\log k_2 = \log A - \frac{Ea}{2.303RT} \tag{3.9}$$

where E_a is the Arrhenius activation energy (kJ/mol), A the Arrhenius factor, R the gas constant (8.314 J/mol K), and T is the solution absolute temperature. When $\log k_2$ is plotted versus $1/T$, a straight line with slope $-E_a/R$ is obtained. The magnitude of activation energy gives an idea about whether the type of adsorption is physical or chemical. The physisorption usually has energies in the range of 5–40 KJ/mol, while higher activation energies (40–800 KJ/mol) suggest involvement of chemisorptions [100].

Table 3.19 Thermodynamic parameters of chromium (VI) and nickel (II) adsorption on modified silica sand.

Adsorbate	Temperature (°C)	ΔG^0 (kJ/mol)	Ea (kJ/mol)	ΔH^0 (kJ/mol)	ΔS^0 (kJ/mol K)
Cr (VI)	25°C	−3.67	79.13	−68.74	−0.243
	30°C	−4.88			
	35°C	-6.10			
Ni (II)	25°C	−5.19	−113.97	−92.35	−0.327
	30°C	−6.82			
	35°C	−8.46			

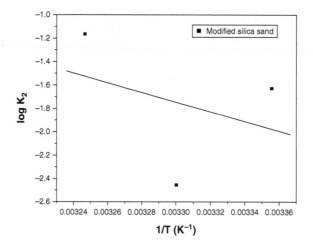

Figure 3.32 Plots of log K_2 versus $1/T$ for adsorption of chromium.

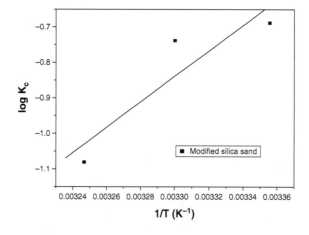

Figure 3.33 Plots of log K_c versus $1/T$ for adsorption of chromium.

The thermodynamic parameters such as change in standard free energy (ΔG^0), enthalpy (ΔH^0), and entropy (ΔS^0) were determined by using the following equations:

$$\log K_c = \frac{\Delta S^0}{2.303R} - \frac{\Delta H^0}{2.303RT} \tag{3.10}$$

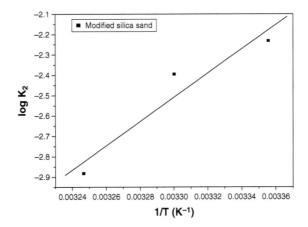

Figure 3.34 Plots of log K_2 versus 1/T for adsorption of nickel.

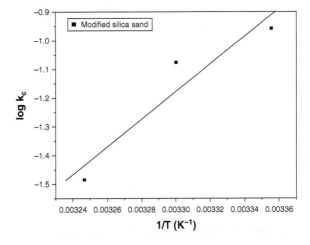

Figure 3.35 Plots of log K_c versus 1/T for adsorption of nickel.

$$\Delta G_{ads} = \Delta H_{ads} - T\Delta S_{ads} \tag{3.11}$$

where R (8.314J/molK) is the gas constant, T (K) the absolute temperature, and k_c (L/g) is the standard thermodynamic equilibrium constant defined by q_e/C_e. Thermodynamic parameters of chromium (VI) and nickel (II) adsorption on silica sand are shown in Table 3.19. By plotting a graph of $\log k_c$ versus $1/T$ (Figures 3.33 and 3.35) the values ΔH^0 and ΔS^0 are then estimated from the slopes and intercepts. The values of activation energy were calculated from the slopes of the graphs in Figures 3.32 and 3.34.

Table 3.19 shows the negative values of ΔG^0 and ΔH^0. This shows that the reaction is spontaneous and it is an exothermic reaction. The negative value of ΔS^0 suggests that there is decrease in randomness at the solid/solution interface occur in the internal structure of the adsorption of Cr (VI) and Ni (II) onto modified silica sand.

4

Conclusions

The experiments were done on removal of Cr (VI) and Ni (II), by using modified silica sand as an adsorbent. The results obtained are satisfactory and show that the removal of these heavy metals was much more dependent on the pH of the wastewater.

The characterization of the silica sand sample reveals that it contains a higher percentage of silica, and it could be in the form of oxides. FTIR shows the presence of silica and 4-membered cyclic compound on the surface. A report based on scanning with electron microscopy revealed that after the treatment of silica sand with 40% H_2SO_4 the roughness of the surface further increases, and might also lead to protonation of the surface of silica sand.

In the experimental study, the removal of Cr (VI) and Ni (II) increases with an increase in contact time, but up to a certain time interval after that, the removal becomes constant. For Cr (VI) equilibrium time is 80 min and for Ni (II) it is found to be at 120 min. Removal of Cr (VI) is maximum at pH 2.5

while for Ni (II) the removal is maximum at pH 6.0. The removal favours low temperature and the maximum removal is obtained at 25°C. The removal % obtained is 80.40% for Cr (VI) and 68.76% for Ni (II).

When dosage of silica sand was varied from 1gm/50ml to 2mg/50ml, the maximum removal was obtained at a high dosage of adsorbent but up to a certain limit, i.e., only up to 1.5gm/50ml; after that removal decreases. This indicates that when the dosage was increased, the adsorption sites were increased and more adsorption takes place. However, at a high dosage of adsorbent, unsaturation of sites takes place and the amount adsorbed decreases. This trend occurs in both cases.

The study of adsorption kinetics shows that the adsorption process follows the pseudo-first-order kinetic model. The thermodynamic study concluded that the adsorption reaction is exothermic and favors low temperatures and is spontaneous in nature.

The conclusion can be summarized as follows:

- The EDAX spectra shows presence of silica in sand sample, it might be in the form of quartz (SiO_2).
- FTIR shows the presence of silica (Si-O) and 4-membered-cyclic compound on the surface of silica sand.
- SEM report reveals that after treatment of silica sand with H_2SO_4, the surface morphology changed due to an increase in surface roughness.
- The batch adsorption experiment shows that the removal of Cr (VI) and Ni (II) increased with an increase in contact time, but only up to a certain time interval, and after that the removal becomes constant. For Cr (VI) equilibrium time is 80 min and for Ni (II) it is found to be at 120 min.
- Results indicate that removal percentage of Cr (VI) and Ni (II) on silica sand decreased at high

solution concentrations compared to the relative values at low solution concentration.

- When dosage of silica sand was varied from 1gm/50ml to 2gm/50ml, the maximum removal was obtained at at high dosage of adsorbent but only up to a certain limit, i.e., only up to 1.5gm/50ml, and after that removal decreases.
- The batch adsorption experiment shows that up to 80.88% of Cr (VI) was removed from aqueous solution at pH 2.5, 25°C, rpm of 150 at silica sand dosage of 1.5gm/50ml.
- For Ni (II) up to 78.72% removal was obtained at pH 6.0, 25°C, rpm of 150 at silica sand dosage of 1.5gm/50ml.
- The pH variation study shows that the adsorption process is highly pH dependent. The removal of Cr (VI) is maximum (80.40%) at pH 2.5, and for Ni (II) the removal is maximum (68.76%) at pH 6.0.
- The adsorption process favors low temperature and the maximum removal is obtained at 25°C, the removal % obtained is 80.40% for Cr (VI) and 68.76% for Ni (II).
- Thermodynamics parameter namely change in free energy (ΔG^0), enthalpy (ΔH^0) and entropy (ΔS^0), were calculated and were found to be –3.67 kJ/mol, –68.74 kJ/mol, and –0.243 kJ/mol K respectively at 25°C for Cr (VI) adsorption process.
- For Ni (II) adsorption process, the thermodynamics parameter namely change in free energy (ΔG^0), enthalpy (ΔH^0), and entropy (ΔS^0), were calculated and were found to be –5.19 kJ/mol, –92.35 kJ/mol, and –0.327 kJ/mol K respectively at 25°C.
- The negative value of ΔG^0 and ΔH^0 shows the spontaneous and exothermic nature of the adsorption process respectively.
- The negative value of ΔS^0 suggest a decrease in randomness at solid/solute interface.

- Activation energy (E_a) was found to be +79.13 KJ/mol in the case of Cr (VI) adsorption process which suggests a chemisorption process, while in the case of Ni (II) adsorption process it is –113.97 KJ/mol which suggests that the process of adsorption is physiosorption.
- The kinetic study shows that adsorption reaction follows pseudo-first order.

References

1. Todd, D.K. (1970). *The Water Encyclopedia*, Water Information Centre, N.Y.
2. Kadirvelu, K., Thamaraiselvi, K., Namasivayam, C. (2001). Adsorption of nickel from aqueous solutions onto activated carbon prepared from coir pith, *Sep. Purif. Tech.*, 24, 497–505.
3. Williams, C.J., Aderhold, D., Edyvean G.J. (1998). Comparison between biosorbent for the removal of metal ions from aqueous solutions, *Water Res.*, 32, 216–224.
4. Wan-Nagh, W.S., Hanafiah, M.A.K.M. (2008). Removal of heavy metal ions from waste water by chemically modified plant waste as adsorbents: a review, *Biores. Tech.*, 99, 3935–3948.
5. Selvaraj, K., Manonmani, S., Pattabhi, S. (2003). Removal of hexavalent chromium using distillery sludge, *Biores. Technol.*, 89,207–211.
6. Mohan, D., Pittman, Jr., C.U. (2006). Activated carbons and low cost adsorbents for remediation of tri and hexavalent chromium from water, *J. Hazardous Mat.*, 137,762–811.
7. Ho, Y.S., Wase, D.A.J., Forster, C.F. (1996). Kinetic studies of competitive heavy metal adsorption by sphagnum moss peat, *Water Res.*, 17,71–74.
8. Hu, J., Chen, G., Lo, I.C.M. (2006). Selective removal of heavy metals from industrial waste water using maghemite nanoparticles, Performance and Mechanism, *J. Environ. Engg.*, ASCE, 709–715.
9. Malkoc, E. (2006). Ni(II) removal from aqueous solutions using cone biomass of *Thuja orientalis*, *J. Hazardous Mat.*, 137, 899–908.
10. USEPA Office of Water (1991). Fact Sheet, National Primary Drinking Water Standards, EPA570/991–012FS.

11. Bielicka, A., Bojanowska, I., Wisniewski, A. (2005). Two faces of chromium-pollutant and bioelement, *Polish Journal of Environmental Studies*, 14 I, 5–10.
12. Alvarez-Ayuso, E., Garcia-Sanchez, A. (2003). Removal of heavy metals from waste water by vermiculites, *Environmental Technology*, 24, 615–625.
13. Buffle, J. (1990). *Complexation Reactions in Aquatic Systems: An Analytical Approach.* Ellis Horwood, New York.
14. Jauhar, S.P. (2002). *Modern's ABC of Chemistry*, Edition, 351.
15. Bodek, I. (1988). *Environmental Inorganic Chemistry: Properties, Processes and Estimation Methods.* New York: Pergamon Press 1988, 7.6–1.
16. Nriagu, J.O. (1988). A silent epidemic of metal poisoning, *Environ. Pollut.* 50, 139–161.
17. Krishna Murti, C.R., P. Viswanathan, P. (1990). *Toxic Metals in the Indian Environment.* New Delhi: Tata McGraw-Hill Publishing Co. Ltd.
18. Helsen, L., Bulck, V.D. (2000). Metal behavior during the low temperature pyrolysis of chromate copper arsenate treated wood waste, *Environ. Sci. Techno.* 34, 2931–2938.
19. Shankar, A.K., Cervantes, C., Loza-Tavera, H., Avudainayagam, S. (2005). Chromium toxicity in plants, *Environ. Int.* 31, 739–753.
20. Park, S., Jung, W.Y. (2001). Removal of chromium by activated carbon fibers plated with copper metal, *Carbon Sci.* 2, 15–21.
21. Mukherjee, A.G. (1986). *Environmental Pollution and Health Hazards, Causes and Control.* New Delhi: Galgotia Pub.
22. Sharma, Y.C., Weng, C.H. (2007). Removal of chromium (VI) from water and wastewater by using riverbed sand: Kinetic and equilibrium studies, *Journal of Hazardous Materials*, 142, 449–454.
23. Marques, M.J., Salvador, A., Morales Rubio, A., de la Guardia, M. (2000). Chromium speciation in liquid matrices: A survey of the literature, *Fresenius J. Anal. Chem.*, 367, 601–613.
24. http://www.Oehha.ca.gov/water/phg/.cal/EPA. Office of the Environmental Health Hazard Assessment, 2001.
25. EPA (1990). Environmental Pollution Control Alternatives, Cincinnati, USA, EPA, 625/5–90/025.
26. Donmez, G., Kocberber, N. (2005). Bioaccumulation of hexavalent chromium by enriched microbial cultures obtained from molasses and NaCl containing media, *Process Biochem.* 40, 2493–2498.

27. Lagergren, S. (1898). Zur theories der sogenannten adsorption geloster stoffee Kung liga svenska vetenskapsakademiens, *Handlingar* 24, 1–39.
28. D.W. Blowes, C.J. Ptacek, J.L. Jambor (1997). In-situ remediation of Cr (VI)- contaminated groundwater using permeable reactive walls: Laboratory studies, *Environ. Sci. Technol.* 31, 3348–3357.
29. Chen, J.M., Hao, O.J. (1998). Microbial Chromium (VI) reduction, Crit. Rev. *Environ. Sci. Technol.* 28, 219–251.
30. Eary, L.E., Rai, D. (1988). Chromate removal for aqueous wastes by reduction with ferrous ion, *Environ. Sci. Technol.* 22, 972–977.
31. Khamis, M., Jumean, Abdo, F.N. (2009). Speciation and removal of chromium from aqueous solution by white, yellow and red UAE sand, *Journal of Hazardous Materials* 169,948–952.
32. Juang, R.S., Shiau, R.C. (2000). Metal removal from aqueous solutions using chitosan enhanced membrane filtration, *J. Membr. Sci,* 21,1091–1097.
33. Yan, G., Viraraghavan, T. (2004). Heavy metal removal in a biosorption column by immobilized M. rouxii biomass, *Bioresour. Technol.* 78,243–249.
34. Kapoor, A., Viraraghavan, T. (1998). Removal of heavy metals from aqueous solution using immobilized fungal biomass in continuous mode, *Water Res.* 32,1968–1967.
35. Wu, D., Sui, Y., He, S., Wang, X., C. Li, Kong, H. (2008). Removal of trivalent chromium from aqueous solution by zeolite synthesized from coal fly ash, *Journal of Hazardous Materials,* 115,415–423.
36. Kumar, P.A., Manabendra, R., Chakraborty, S. (2007). Hexavalent chromium removal from waste water using aniline formaldehyde condensate coated silica gel, *J Hazard. Materials.* 143,24–32.
37. Alvarez-Ayuso, E., Garcia-Sanchez, A., Querol, X. (2007). Adsorption of Cr (VI) from synthetic solutions and electroplating wastewater on amorphous aluminum oxide, *J. Hazard. Mater.* 142,191–198.
38. Malkoc, E., Nuhoglu, Y. (2005). Investigation of nickel(II) removal from aqueous solutions using tea factory waste, *J. Hazardous Mat.* 127,120–128.
39. Cimino, G., Caristi, C. (1990). Acute toxicity of heavy metals to aerobic digestion of waste cheese whey, *Biol. Wastes,* 33,201–210.

40. Madoni, P., Davoli, D., Gorbi, G., Vescovi, L. (1996). Toxic effects of heavy metals on the activated sludge protozoan community, Water Res., 30,135–142.
41. Bodek, I. (1988). *Environmental Inorganic Chemistry: Properties, Processes and Estimation Methods.* New York: Pergamon Press, 7.11–1.
42. Murti, C.R.K., Viswanathan, P. (1990). *Toxic Metals in the Indian Environment.* New Delhi: Tata McGraw-Hill Publishing Co. Ltd., 149,162–164.
43. Eary, L.E., Rai, D. (1988). Chromate removal for aqueous wastes by reduction with ferrous ion, *Environ. Sci. Technol.* 22, 972–977.
44. Levankumar, L., Muthukumaran, V., Gobinath, M.B. (2009). Batch adsorption and kinetics of Cr (VI) removal from aqueous solutions by Ocimum americanum L. seed pods, *Journal of Hazardous Materials* 161,709–713.
45. Yan, G., Viraraghavan, T. (2004). Heavy metal removal in a biosorption column by immobilized M. rouxii biomass, *Bioresour. Technol.* 78(3), 243–249.
46. Mukherjee, A.G. (1986). *Environmental Pollution and Health Hazards: Causes and Control.* New Delhi: Galgotia Pub.
47. Nriagu, J.O. (1988). A silent epidemic of metal poisoning, *Environmental Pollut.* 50,139–161.
48. Kanan, K. (1985). *Fundamentals of Environmental Pollution, 1st Edition.* S. Chand & Co., New Delhi, India.
49. Förstner, U., Wittmann, G.T.W. (1985). *Metals in Aquatic Environment.* Springer-Verlag, New York.
50. Arslan, G., Pehlivan, E. (2007). Batch removal of Cr(VI) from aqueous solution by Turkish brown coal, *Biores. Technol.*, 98, 2836–2845.
51. Boujelben, N., Bouzid, J., Elouear, Z. (2009). Adsorption of nickel and copper onto natural iron oxide-coated sand from aqueous solutions: Study in single and binary systems, *Journal of Hazardous Materials* 163, 376–382.
52. Sawyer, C.N., McCarty, P.L., Parkin, G.F. *Chemistry for Environmental Engineering and Science, 5th Edition.* New Delhi: Tata McGraw-Hill Publishing Co. Ltd., 717.
53. Barati, A.H., Maleki, A., Alasvand, M. (2010). Multi-trace elements level in drinking water and the prevalence of multi-chronic arsenical poisoning in residents in the west area of Iran, *Science of The Total Environment*, 408,1523–1529.

54. Sharma, Y.C. (2003). Cr (VI) removal from industrial effluents by adsorption on an indigenous low-cost material, Colloids and Surfaces A, *Physiochem. Eng. Aspects*, 215,155–162.
55. Sharma, Y.C., Uma, Srivastava, V., Srivastava, J., Mahto, M. (2007). Reclamation of Cr (VI) rich water and wastewater by wollastonite, *Chemical Engineering Journal* 127,151–156.
56. Babel, S, Kurniawan, T.A. (2003). Low-cost adsorbents for heavy metals uptake from contaminated water: A review, *Journal of Hazardous Mat.* B79, 219–243.
57. Yan, G., Viraraghavan, T. (2004). Heavy metal removal in a biosorption column by immobilized M. rouxii biomass, *Bioresour. Technol.* 78, 243–249.
58. Mor, S., Khaiwal, R., Bishnoi, N.R. (2007). Adsorption of chromium from aqueous solution by activated alumina and activated charcoal, *Bioresource Technology,* 98, 954–957.
59. Sharma, Y.C., Singh, B., Agrawal, A., Weng, C.H. (2008). Removal of chromium by riverbed sand from water and wastewater; Effect of important parameters, *Journal of Hazardous Materials,* 151,789–793.
60. Lokeshwari, N., Keshava Joshi (2009). Biosorption of heavy metal (chromium) using biomass, *Global Journal of Environmental Research* 3, 29–35.
61. Baral, S.S., Das, S.N., Rath, P. (2006). Hexavalent chromium removal from aqueous solution by adsorption on treated saw dust, *Biochemical Engineering Journals*, 31, 216–222.
62. Levankumar, L., Muthukumaran, V., Gobinath, M.B. (2009). Batch adsorption and kinetics of Cr (VI) removal from aqueous solutions by Ocimum americanum L. seed pods, *Journal of Hazardous Materials* 161,709–713.
63. Eary, L.E., Rai, D. (1988). Chromate removal for aqueous wastes by reduction with ferrous ion, *Environ. Sci. Technol.* 22, 972–977.
64. Ayuso, E.A., Sanchez, A.G. (2003). Removal of heavy metals from waste water by vermiculites, *Environmental Technology,* 24, 615–625.
65. Petruzzelli, G., Petronio, B.M., Gennaro, M.C., Vanni, A., Lubrano, L., Liberatori, A. (1992). Effect of a sewage sludge extract on the sorption process of cadmium and nickel by soil. *Environmental Technology,* 13,1023–1032.
66. Poulsen, I.F., Bruun Hansen, H.C. (2000). Soil sorption of nickel in presence of citrate or arginine. *Water, Air and Soil Pollution* 120, 249–259.

67. Santos Yabe, M.J., de Oliveira, E. (2003). Heavy metals removal in industrial effluents by sequential adsorbent treatment. *Advances in Environmental Research* 7, 263–272.

68. Boujelben, N., Bouzid, J., Elouear, Z. (2009). Adsorption of nickel and copper onto natural iron oxide-coated sand from aqueous solutions: Study in single and binary systems, *Journal of Hazardous Materials* 163, 376–382.

69. Choksi, P.M., Joshi, V.Y. (2007). Adsorption kinetic study for removal of nickel (II) and aluminium (III) from aqueous solution by natural adsorbents. *Desalination* 208, 216–231.

70. Shukla, S.S., Yu, L.J., Dorris, K.L., Shukla, A. (2005). Removal of nickel from aqueous solutions by sawdust. *Journal of Hazardous Materials* 121, 243–246.

71. Li, C., Champagne, P. (2009). Fixed-bed column study for removal of cadmium (II) and nickel (II) ions from aqueous solutions using peat and mollusc shells. *Journal of Hazardous Materials* 171, 872–878.

72. Doan, H.D., Lohi, A., Dang, V.B.H., Dang-Vu, T. (2008). Removal of Zn^{+2} and Ni^{+2} by adsorption in a fixed bed of wheat straw. *Process Safety and Environment Protection*, 86, 259–267.

73. Kraemer, E.O. (1930). in *A Treatise on Physical Chemical Chemistry.* H.S. Taylor (Ed.), II Ed., Vol. II. New York: D. Van Nostrand Co., Inc.

74. Giles, C.H., Macewan, T.H., Nakhwa, S.N., Smith, D. (1960). Studies in adsorption, XI. *J. Chem. Soc.*, 3973–3993.

75. *APHA Standard Methods for the Examination of Water and Wastewater* (1985). 14th Ed., Washington.

76. Sincero, A.P., Sincero, G.A. (2006). *Environnmental Engineering: A Design Approach.* Prentice-Hall of India Private Limited, 401.

77. Oura, K., *et al.* (2003). *Surface Science: An Introduction.* Berlin: Springer, ISBN 978-3-540-00545-2.

78. Desjonqueres, M.C., *et al.* (1993). *Concepts in Surface Physics.* Springer-Verlag, ISBN 978-0-387-56506-4.

79. Lüth, H., *et al.* (1993). *Surfaces and Interfaces of Solids.* Springer-Verlag, ISBN 978-3-540-56840-7.

80. Oura, K., Lifshits, V.G., Saranin, A.A., Zotov, A.V., Katayama, M. (2003). *Surface Science: An Introduction.* Berlin: Springer.

81. Hall, K.R., Eagleton, L.C., Acrivos, A., Vermeulen, T. (1996). Pore and solid diffusion kinetics in fixed-bed adsorption under constant pattern conditions, *Ind. Eng. Chem. Fundam.* 5, 212–223.

82. Langmuir, I. (1916). The constitution and fundamental properties of solids and liquids. part i. solids. I. *Journal of the American Chemical Society* 38, 2221–2295.

83. Sharma, A., Bhattacharyya, K.G. (2004). Adsorption of chromium (VI) on *Azadirechta Indica* (Neem) leaf powder, *Adsorption* 10,327–338.

84. Zheng, W., Qian, J.Z., Li-Li, C., Wei, S., Kong Yu (2009). Adsorption of Cr (VI) by attapulgite-zeolite composite ceramisite from aqueous solution, *Res. J. Chem. Environ.*, 13, 23–39.

85. Namasivayam, C., Jeyakumar, R., Yamuna, R.T. (1994). Dye removal from waste water by adsorption on 'waste' Fe (III)/ Cr (III) hydroxide, *Waste Manage.*, 14, 643–648.

86. Lagergren, S., (1898). Zur theorie der sogenannten adsorption geloster stoffle, K. Sven.Vetenskapsakad. *Handl.*, 24, 1–39.

87. McKay, G., Ho, Y.S. (1999). Pseudo-second order model for sorption processes, *Process Biochem.*, 34, 451–465.

88. Demirbas, A. (2005). Adsorption of Cr(III) and Cr(VI) ions from aqueous solutions onto modified lignin, *Energy Sources*, 27,1449–1455.

89. Malkoc, E., Nuhoglu, Y. (2007). Potential of tea factory waste for chromium (VI) removal from aqueous solutions: thermodynamics and kinetic studies, *Sep. Purif. Technol.*, 54,291–298.

90. Gode, F., Pehlivan, E. (2005). Adsorption of Cr(III) by Turkish brown coals, *Fuel Processing Technol.*, 86,875–884.

91. Sergei, I.L., Andrei, I.L., Olga, L.G., Lilia, P.T., Joaquim, V., Isabel, M.F., Svetlana, B.L. (2004). Kinetics and thermodynamics of the Cr(III) adsorption on the activated carbon from co-mingled waste, *Colloids and Surfaces A: Physicochem. And Engg. Aspects*, 242, 151–158.

92. Uysal, M., Ar, I. (2007). Removal of Cr(VI) from industrial wastewaters by adsorption Part I: Determination of optimum conditions, *J. Hazardous Mater.*

93. Gonzalez, J.R., Videa, J.R.P., Rodriguez, E., Ramirez, S.L., Torresdey, J.L.G. (2004). Determination of thermodynamic parameters of Cr(VI) adsorption from aqueous solutions onto *Agave lechuguilla* biomass, *J. Chem. Thermodynamics*, 37, 343–347.

94. Oguz, E. (2005). Adsorption characteristics and kinetics of the Cr(VI) on the *Thuja orientalis*, *Colloids and Surfaces A: Physicochem. And Engg. Aspects*, 252,121–128.

95. Senthil Kumar, P., Ramalingam, S., Dinesh Kirupha, S., A. Murugesan, A., T. Vidhyadevi, T., Sivanesan, S. (2011). Adsorption

behavior of nickel (II) onto cashew nut shell: Equilibrium, thermodynamics, kinetics, mechanism and process design, *Chemical Engineering Journal*, 167,122–131.

96. Argun, M.E., (2008). Use of clinoptilolite for the removal of nickel ions from water: Kinetics and thermodynamics, 150,587–595.
97. Debnath, S., Ghosh, U.C. (2009). Nanostructured hydrous titanium(IV) oxide: Synthesis, characterization and Ni(II) adsorption behavior, *Chemical Engineering Journal*, 152,480–491.
98. Kandah, M.I., Meunier, J.L. (2007). Removal of nickel ions from water by multi-walled carbon nanotubes, 146,283–288.
99. Hameed, B.H., Ahmad, A.A., Aziz, N. (2007). Isotherms, kinetics and thermodynamics of acid dye adsorption on activated palm ash, *Chemical Engg. Journal*, 133,195–203.
100. Nollet, H., Roels, M., Lutgen, P., Van der Meeren, P., Verstraete, W. (2003). Removal of PCBs from wastewater using fly ash, *Chemosphere*, 53,655–665.

Index

Also of Interest

Check out these published and forthcoming related titles from Scrivener Publishing

Green Chemistry and Environmental Remediation
Edited by Rashmi Sanghi and Vandana Singh
December 2011. ISBN 978-0-470-94308-3

Bioremediation and Sustainability: Research and Applications
Edited by Romeela Mohee and Ackmez Mudhoo
June 2012. ISBN 978-1-1180-6284-5

Pretreatment for Enhanced Anaerobic Technology
Edited by Ackmez Mudhoo
June 2012. ISBN 978-1-1180-6285-2

Reverse Osmosis: Industrial Applications and Processes
By Jane Kucera
Published 2010. ISBN 978-0-470-61843-1

Electrochemical Water Processing
By Ralph Zito
Published 2011. ISBN 978-1-118-09871-4